AF594938

New Trends and Applications in Femtosecond Laser Micromachining

New Trends and Applications in Femtosecond Laser Micromachining

Editors

Rebeca Martínez Vázquez
Francesca Bragheri
Petra Paiè

MDPI • Basel • Beijing • Wuhan • Barcelona • Belgrade • Manchester • Tokyo • Cluj • Tianjin

Editors
Rebeca Martínez Vázquez
Intitute for Photonics and Nanotechnologies
CNR
Milan
Italy

Francesca Bragheri
Institute for Photonics and Nanotechnologies
CNR
Milan
Italy

Petra Paiè
Physics Department
Politecnico di Milano
Milan
Italy

Editorial Office
MDPI
St. Alban-Anlage 66
4052 Basel, Switzerland

This is a reprint of articles from the Special Issue published online in the open access journal *Micromachines* (ISSN 2072-666X) (available at: www.mdpi.com/journal/micromachines/special_issues/Laser_Micromachining).

For citation purposes, cite each article independently as indicated on the article page online and as indicated below:

LastName, A.A.; LastName, B.B.; LastName, C.C. Article Title. *Journal Name* **Year**, *Volume Number*, Page Range.

ISBN 978-3-0365-3310-0 (Hbk)
ISBN 978-3-0365-3309-4 (PDF)

Contents

Editorial

Editorial for the Special Issue on New Trends and Applications in Femtosecond Laser Micromachining

Francesca Bragheri , Petra Paiè and Rebeca Martínez Vázquez *

Istituto di Fotonica e Nanotecnologie (IFN)—CNR, Piazza L. da Vinci 32, 20133 Milano, Italy; francesca.bragheri@ifn.cnr.it (F.B.); petra.paie@polimi.it (P.P.)
* Correspondence: rebeca.martinez@polimi.it

Femtosecond laser micromachining is becoming an established fabrication technique for transparent material processing in three dimensions [1]. Laser writing permits obtaining different material modifications, depending on the laser parameters as well as on the material properties. For instance, it allows creating micrometer-size structures by two-photon polymerization (additive approach), as well as creating empty channels inside the material (subtractive approach). On the other hand, transformative approaches are also permitted, thus locally modifying the material properties for optical waveguide fabrication and material welding. Therefore, the capabilities of femtosecond laser micromachining open the doors to a plethora of applications ranging from the biological to the information technology field.

This Special Issue is composed of 10 contributions, including original research and reviews. The review paper from Butkute et al. [2] reports a brief but exhaustive overview of the significant advancements in the femtosecond laser machining of glasses, reviewing the possible modifications of the material and highlighting application examples such as optical waveguiding and microfluidic systems. Following the unique capabilities of femtosecond laser technologies, we highlight four macro areas of applications in which we classified the contributions of the issue.

Material processing. In this category, we include papers that discuss the use of femtosecond lasers either to structure or to process the substrates of interest. Most of the previously mentioned approaches can be exploited for the precise material structuring. Umenne [3] exploits the femtosecond laser ablation of YBCO thin films to fabricate sub-micron and nano-sized bridges to be used as Josephson junctions. A new irradiation method based on a double pulse approach has been proposed by Stankevic et al. [4] to optimize the realization of hollow structures in bulk fused silica. The irradiation with two orthogonally polarized laser beams leads to the formation of nanogratings with grid-like morphology, giving rise to an improvement of chemical etching anisotropy and to a faster processing speed. Microfractures produced by femtosecond (FS) lasers have been exploited by Gaudiuso et al. [5] to achieve stealth dicing of quartz with neat and flat cut edges in a single pass. The result is ascribed to tensile stresses that after relaxation produce microfractures that are guided by the laser to achieve cutting.

Lab-on-chip. Here, we include papers that demonstrate the potential of the use of femtosecond lasers in the bio-chemical fields. Microfluidic platforms are widely used and can be realized by diverse technologies either in plastic or in glass. Anyway, femtosecond laser micromachining still plays a key role for its intrinsic three-dimensional capabilities as demonstrated by Qi et al. [6] who realized an efficient 3D mixer based on Baker's transformation principle that could find applications in microfluidic synthesis of materials or fine chemistry microreactions. The device is realized by femtosecond irradiation followed by chemical etching and hydroxide-catalysis bonding in fused silica. Different applications are envisaged for the reservoir etched in borosilicate glass by high-fluence FS lasers by Owusu-Ansah [7] et al. In their work, they optimize the etching depth as a function of

Citation: Bragheri, F.; Paiè, P.; Vázquez, R.M. Editorial for the Special Issue on New Trends and Applications in Femtosecond Laser Micromachining. *Micromachines* **2022**, *13*, 150. https://doi.org/10.3390/mi13020150

Received: 13 January 2022
Accepted: 15 January 2022
Published: 19 January 2022

fluence and fabricate multi-depth reservoirs with large areas for possible application in microfluidic transport investigation or CO_2 storage.

Biophotonics. In this category, we decided to include papers where femtosecond lasers are used for machining or analysis specifically in the biomedical field. Keleman et al. [8] exploited two-photon polymerization for the fabrication of microtools that thanks to their three-dimensional shape could be easily actuated by optical tweezer and exploited to probe single cell elasticity, without inducing photodamage to the sample. FS-laser technology in the medical field is used also as precise and reliable tool, for example, in ophthalmic surgery. As reported in the review by Latz et al. [9] some of the applications in this field still require research and development, while operations such as FS-laser-assisted cataract and corneal surgery have reached highly standardized levels worldwide.

Integrated optics. In this category, we include papers where laser modifications are exploited to fabricate optical waveguides or other integrated optical components. A very active application field is the one concerning quantum optics; in this contest, the paper by Li et al. [10] reports a photonic quantum chip fabricated in borosilicate glass by femtosecond laser transformative approach. The high quality of the chip encompassing a Hadamard and a CNOT gate to generate a four path-encoded Bell states is demonstrated by the high value of the fidelity of the reconstructed truth table. Femtosecond laser irradiation followed by chemical etching is instead used by Sala et al. [11] for the fabrication of micromirrors that could be included in integrated photonic or optofluidic chips. The effects of thermal annealing in reducing the residual roughness are evaluated with a profilometer to optimize the smoothening performed by the oven.

We would like to thank all the contributors for submitting their papers to this Special Issue. We also thank all the reviewers for dedicating their time to help improve the quality of the submitted papers.

Funding: F.B. and P.P. acknowledge funding from EU under the European Union's Horizon2020 FET Open program (PROCHIP project - grant agreement no.801336). R.M.V. acknowledges funding from EU the European Union's Horizon2020 FET Open program (X-PIC project-grant agreement no. 964588).

Conflicts of Interest: The authors declare no conflict of interest.

References

1. Osellame, R.; Cerullo, G.; Ramponi, R. *Femtosecond Laser Micromachining: Photonic and Microfluidic Devices in Transparent Materials*; Springer Science and Business Media: New York, NY, USA, 2012.
2. Butkute, A.; Jonusauskas, L. 3D Manufacturing of Glass Microstructures Using Femtosecond Laser. *Micromachines* **2021**, *12*, 499. [CrossRef] [PubMed]
3. Umenne, P. AFM Analysis of Micron and Sub-Micron Sized Bridges Fabricated Using the Femtosecond Laser on YBCO Thin Films. *Micromachines* **2020**, *11*, 1088. [CrossRef] [PubMed]
4. Stankevic, V.; Karosas, J.; Raciukaitis, G.; Gecys, P. Improvement of Etching Anisotropy in Fused Silica by Double-Pulse Fabrication. *Micromachines* **2020**, *11*, 483. [CrossRef] [PubMed]
5. Gaudiuso, C.; Volpe, A.; Ancona, A. One-Step Femtosecond Laser Stealth Dicing of Quartz. *Micromachines* **2020**, *11*, 327. [CrossRef] [PubMed]
6. Qi, J.; Li, W.; Chu, W.; Yu, J.; Wu, M.; Liang, Y.; Yin, D.; Wang, P.; Wang, Z.; Wang, M.; et al. A Microfluidic Mixer of High Throughput Fabricated in Glass Using Femtosecond Laser Micromachining Combined with Glass Bonding. *Micromachines* **2020**, *11*, 213. [CrossRef] [PubMed]
7. Owusu-Ansah, E.; Dalton, C. Fabrication of a 3D Multi-Depth Reservoir Micromodel in Borosilicate Glass Using Femtosecond Laser Material Processing. *Micromachines* **2020**, *11*, 1082. [CrossRef] [PubMed]
8. Grexa, I.; Fekete, T.; Molnar, J.; Molnar, K.; Vizsnyczai, G.; Ormos, P.; Kelemen, L. Single-Cell Elasticity Measurement with an Optically Actuated Microrobot. *Micromachines* **2020**, *11*, 882. [CrossRef] [PubMed]
9. Latz, C.; Asshauer, T.; Rathjen, C.; Mirshahi, A. Femtosecond-Laser Assisted Surgery of the Eye: Overview and Impact of the Low-Energy Concept. *Micromachines* **2021**, *12*, 122. [CrossRef] [PubMed]
10. Li, M.; Zhang, Q.; Chen, Y.; Ren, X.; Gong, Q.; Li, Y. Femtosecond Laser Direct Writing of Integrated Photonic Quantum Chips for Generating Path-Encoded Bell States. *Micromachines* **2020**, *11*, 1111. [CrossRef] [PubMed]
11. Sala, F.; Paiè, P.; Martínez Vázquez, R.; Osellame, R.; Bragheri, F. Effects of Thermal Annealing on Femtosecond Laser Micromachined Glass Surfaces. *Micromachines* **2021**, *12*, 180. [CrossRef] [PubMed]

Review

3D Manufacturing of Glass Microstructures Using Femtosecond Laser

Agnė Butkutė [1,2,*] and Linas Jonušauskas [1,2,*]

[1] Femtika Ltd., Saulėtekio Ave. 15, LT-10224 Vilnius, Lithuania
[2] Laser Research Center, Vilnius University, Saulėtekio Ave. 10, LT-10223 Vilnius, Lithuania
* Correspondence: agne@femtika.lt (A.B.); linas@femtika.lt (L.J.)

Abstract: The rapid expansion of femtosecond (fs) laser technology brought previously unavailable capabilities to laser material processing. One of the areas which benefited the most due to these advances was the 3D processing of transparent dielectrics, namely glasses and crystals. This review is dedicated to overviewing the significant advances in the field. First, the underlying physical mechanism of material interaction with ultrashort pulses is discussed, highlighting how it can be exploited for volumetric, high-precision 3D processing. Next, three distinct transparent material modification types are introduced, fundamental differences between them are explained, possible applications are highlighted. It is shown that, due to the flexibility of fs pulse fabrication, an array of structures can be produced, starting with nanophotonic elements like integrated waveguides and photonic crystals, ending with a cm-scale microfluidic system with micro-precision integrated elements. Possible limitations to each processing regime as well as how these could be overcome are discussed. Further directions for the field development are highlighted, taking into account how it could synergize with other fs-laser-based manufacturing techniques.

Keywords: femtosecond laser; glass micromachining; 3D structuring

Citation: Butkutė, A.; Jonušauskas, L. 3D Manufacturing of Glass Microstructures Using Femtosecond Laser. *Micromachines* **2021**, *12*, 499. https://doi.org/10.3390/mi12050499

Academic Editors: Rebeca Martínez Vázquez, Francesca Bragheri and Petra Paiè

Received: 28 February 2021
Accepted: 12 April 2021
Published: 28 April 2021

1. Introduction

Glass and related transparent dielectrics play an important role in huge variety of applications. Through the millennia it found its way into all aspects of life, starting from construction engineering and building expanding to nanotechnology and space exploration. Glass owns its applicability to a combination of superb properties, such as being mechanically robust, transparent to the visible part of the electromagnetic spectrum, chemically inert in organic solvents and biocompatible. While glasses are not as diverse as metals [1,2] or polymers [3,4], they also can be modified in various ways [5–8] expanding their applicability. Nevertheless, ways to process glass are surprisingly limited, mostly relying on various kinds of casting [9,10] or mechanical cutting [11,12]. Due to the rapid development of science and technology, the need for more advanced and, most importantly, precise and flexible ways to produce glass structures arose. To accommodate the needs for nanotechnology various chemical deposition [13,14] or wet etching [15,16] techniques were employed. At the same time, direct [17,18] and indirect [19,20] additive 3D manufacturing of glass was demonstrated. However, in all of the highlighted processes, severe drawbacks and compromises are limiting their applicability. A more diverse and flexible approach is highly sought after.

One of the premier candidates for the best possible blend between free-form fabrication capability, flexibility, and throughput is the usage of femtosecond (fs) laser [21]. Indeed, it was already employed in a huge variety of fields [22]. The primary advantage of using ultrashort pulses is the possibility to control the thermal aspects of the light-matter interaction [23–25]. Furthermore, interaction can be nonlinear, meaning that materials transparent to the incident wavelength can be processed [26]. Combined, these two aspects make fs pulses extremely attractive for glass processing. Thus, this review is dedicated to highlighting how fs pulses can be employed for free-form 3D microstructuring of glass. Dielectric crystals will also be mentioned where relevant, as a lot of glass-related processing techniques can be applied for them as well. First, the physical phenomena of ultrafast light-matter interaction will be discussed, explaining how it can be tuned and localized. Then, various processing techniques based on these interactions will be explained. These include waveguide and other photonic element inscribing, zone plates, selective glass etching (SLE), and ablation. Primary attention will be given to real-life applications of shown techniques as well as connections with other manufacturing methodologies. Combined, this will provide a comprehensive outlook to the current progress of fs glass fabrication and highlight the directions where the field is headed.

2. Ultrafast Radiation and Material Interaction

2.1. The Nature of Nonlinear Radiation and Material Interaction

To give a proper background to the technologies that will be discussed further in the review, we will begin by explaining the nature of ultrafast light and material interaction. When the material is processed by using fs pulses radiation the material is affected by very high peak intensity I (more than GW/cm^2). As I is the ratio between the laser power P and focused spot size with a radius of ω_0, the two main ways to raise it are an increase in P or focus a laser beam to smaller spot size. The first requirement is fulfilled by high peak power (P_P) of fs pulses ($\sim$GW), while small spot sizes are achieved by sharp focusing—high numerical aperture (NA > 0.2) objectives or lenses. Material response to high I is nonlinear and conceptually is different from the material response to low-intensity radiation which is generated by non-coherent light sources. When radiation travels through dielectric material it polarizes material. When radiation intensity is not high enough material polarization linearly depends on radiation electric field E (1). Keeping in mind that $I \sim |E^2|$, when radiation creates an electric field which is stronger than 10^7 V/m, material polarization dependency on the electric field becomes nonlinear (2).

$$P(t) = \epsilon_0 \chi^{(1)} E(t), \tag{1}$$

$$P(t) = \varepsilon_0 \chi^{(1)} E(t) + \varepsilon_0 \chi^{(2)} E^{(2)}(t) + \varepsilon_0 \chi^{(3)} E^{(3)}(t) + \ldots, \tag{2}$$

here ε_0 is vacuum dielectric permeability, $\chi^{(1)}$ is linear optical sensitivity, $\chi^{(2)}$—second order nonlinear optical sensitivity, $\chi^{(3)}$—third order nonlinear optical sensitivity.

Different member in nonlinear polarization equation decides different phenomenon observed in material. It also might exclude some materials from certain interaction regimes. For instance, effects that are decided by square nonlinear sensitivity, like sum and difference frequency generation, are observed only in non-central symmetrical crystals. Meanwhile, cubic nonlinear sensitivity affected phenomena are observed in all-dielectric materials. These effects are multi-photon absorption, tunneling ionization and other radiation self-interaction, such as light thread formation. It is important to note that various nonlinear light-interaction regimes can be induced simultaneously, leading to a quite complex picture of the process. Thus, when processing non-central symmetric materials a second-order interaction have to be taken into account, or how the Kerr effect might change propagation conditions when high I is present [27].

2.2. Nonlinear Effects

As it was mentioned before when the material is affected by high-intensity radiation various nonlinear light and material interaction effect appears which are governed by different order nonlinear sensitivity. One of these effects is multiphoton absorption. Multiphoton absorption is simultaneous two or more photons absorption through virtual levels from the ground to an excited state. Two-photon absorption is decided by second-order nonlinear optical sensitivity. Meanwhile, three-photon absorption is decided by third-order nonlinear optical sensitivity. Hence, higher-order multiphoton absorption is decided by higher-order nonlinear optical sensitivity respectively. Most of the time in nonlinear optics multiphoton optics is a negative effect that distorts radiation. However multiphoton absorption can be used for positive practical applications such as multiphoton spectroscopy, microscopy, and transparent material laser processing.

In simplified terms, to induce a change in the dielectric material the electrons have to be excited from the valance band to the conductive band. Numerous processes might induce it. The most important and the most prevailing nonlinear effects when a material is illuminated with high I fs radiation are multiphoton ionization, tunneling ionization, and avalanche ionization. Multiphoton ionization is a process when an electron absorbs several photons simultaneously and is excited *via* virtual levels. The principal scheme of this process is shown in Figure 1b. Another nonlinear effect is tunneling ionization. Intense laser radiation can distort the potential barrier which holds electrons to material atoms. Because of the potential barrier decrease possibility of electron tunneling increases and in this way more and more electrons are excited to the conduction band. This process is depicted in Figure 1c.

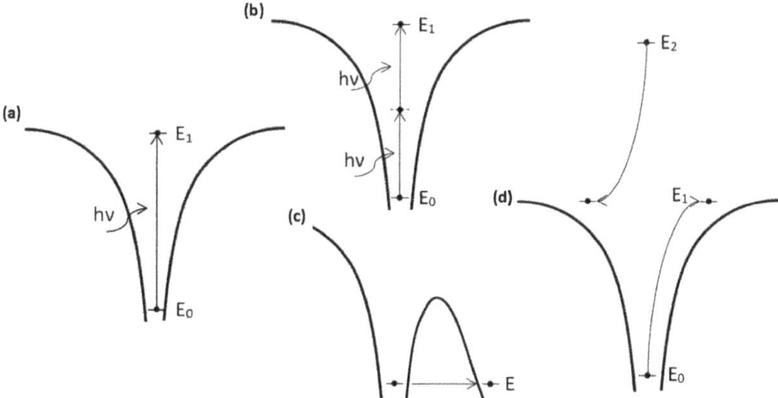

Figure 1. Principle schemes of main nonlinear processes: (**a**) linear ionization, (**b**) nonlinear (multi-photon) ionization, (**c**) tunneling ionization and (**d**) avalanche ionization.

Which process (multi-photon ionization or tunneling ionization) is going to dominate depends on laser ω and I. This dependency is described by Keldish parameter [28] which can be expressed as [29]:

$$\gamma = \frac{\omega}{e}\sqrt{\frac{m_e \Delta E \varepsilon_0}{I}}, \quad (3)$$

here e—electron charge, m—electron mass, c—speed of light, n—linear refraction index, I—laser intensity, ϵ_0—vacuum dielectric penetration, ΔE—reserve strip spacing.

When γ is more than 1.5, multi-photon ionization will dominate, when γ is less than 1.5, tunneling ionization will dominate. Meanwhile, when the Keldish parameter is close to 1.5, transitional regime ionization occurs. In this case, multi-photon and tunneling ionization have approximately the same influence on material ionization. To sum up, tunneling ionization prevails in high I and low ω regimes. In contrast, multi-photon ionization dominates in high I and high ω cases. In laser metrical processing to reach the desired effect, it is necessary to control which effect will dominate. Fortunately, modern amplified fs-laser systems can be easily tuned to cover all the processing regimes.

Another important process that appears because of nonlinear radiation and material interaction is avalanche ionization. The principal scheme of this process is shown in Figure 1d. When radiation interacts with material electrons are released from material atoms. Continuously, irradiating material by intense electric field-free electrons is accelerated and their energy can be enough to release other electrons from neighboring atoms. In this way more and more electrons are released. Finally, the critical density of electrons in a material is reached and the material can be modified [27]. For this process to appear sufficient amount of seed electrons is required. At the same time, to achieve sufficiently high electron densities using avalanche the time for the process to appear have to be sufficiently long (>100 fs). Therefore, this process is more common when longer amplified pulses are used. Also, care should be taken, because if the process is too strong it can become uncontrollable, worsening structuring quality. Thus, care should be taken if this process is present during DLW.

2.3. Thermal Effects

When a material is processed with relatively long laser pulses (>10 ps) thermal effects can be observed in the material. It is the result of heat dissipation from the laser affected area which occurs while a long laser pulse is still interacting with the material in the focal point. This process takes its place around 1 ns after radiation absorption [30] and usually reduces material processing quality. When quantifying thermal effect the term Laser Affected Zone (LAZ) is introduced. It encompasses all the volume affected by the heat generated by a laser, even if it is outside of the focal point [31].

Usage of an ultrafast laser can allow control and if needed greatly reduce thermal effects caused by the light-matter interaction. Looking from the temporal perspective, interactions needed for processing occur rather fast. For instance, material ionization occurs after $\sim$1 ps, followed by ablation and removal after $\sim$100 ps [30]. Therefore, if no additional energy is introduced while subsequent thermal effects take place it can be greatly reduced. As a result in the fs pulses, the case material does not suffer from substantial thermal effects. In this way, LAZ is minimized. That is why femtosecond material laser processing sometimes is called cold material processing. Nevertheless, it is important to note that alongside pulse duration pulse repetition rate also plays important role in the thermal aspect of processing. If several pulses reach the affected zone before initial heat dissipation takes place, even fs pulses can induce material melting and noticeable LAZ. Another important factor for thermal accumulation process is the deposited energy. By maintaining the same pulse duration and pulse repetition rate and at the same time by increasing energy of each pulse the sample is affected by greater radiation dose. If the energy does not have enough time to be relaxed out of sample by increasing pulse energy we obtain higher thermal effects. Higher pulse energy shows greater affected volume around the laser focus [31] that means stronger thermal effects. While in some cases this might be detrimental, it can also be useful in some applications like laser welding. This distinction will be made through the article where it applies.

These thermal effects are especially important for glass processing. Glass is a quite fragile material that can be easily broken by treating it with stresses [32] or temperature caused volume tensions [33]. By processing glass with longer pulses or even CW laser it is very hard to avoid cracking of glass [34]. However, short pulses lasers development allows minimizing thermal effects by using shorter pulse duration [35] and process material locally without affecting surrounding volume.

3. Peculiarities of Glass Processing Using Fs Radiation

Light-based glass processing is not a straightforward process. Most glasses (with some exceptions like chalcogenide glasses) are transparent materials for near-UV, visible and near-IR radiation, meaning that linear interaction with such photons is negligible. As a result, one way to induce a modification in glasses is by choosing a wavelength that can be absorbed by a particular glass. Some works use such interaction for glass surface structuring based on linear glass absorption [36,37]. Such direct removal of the material by laser beam This is called laser ablation. Nevertheless, 3D processing in such cases is greatly limited due to direct absorption being a surface-bound process limiting its use for 3D structuring.

Alternatively, intense and focused radiation can be used. Then interaction becomes nonlinear, resulting in melting and removal of the focal volume. For this reason, most laser-based glass 3D processing techniques are based on such processes. As we already have discussed, when radiation I is high enough (GW/cm^2–PW/cm^2) multiple photons (in most cases two) can be absorbed simultaneously during multi-photon ionisation [27]. Because high enough I for this interaction can only be reached by focused laser light, this creates inherent localization of the process. The scheme of this idea is shown in Figure 2. Keeping in mind, that in most cases NA in the range from 0.01 to 1.4 is used, the volume in which the modification is induced can range from a hundred nm to hundreds on µm in a transverse direction and from a few µm to several mm in a longitudinal direction. The elongated modification is an inherent feature of Gaussian beam normally used for such processing [38] while other types of voxel shapes being possible with beam structuring. Due to the nonlinear nature of these interactions sub-diffraction limited features are also possible, both directly and indirectly. Regardless, localization and selectivity of the process in the volume of glass is a critical component why fs laser is the primary tool for high precision 3D fabrication. Keeping in mind that laser can be used in a multitude of ways and not only for glass processing any laser-based fabrication process can be called direct laser writing DLW [39].

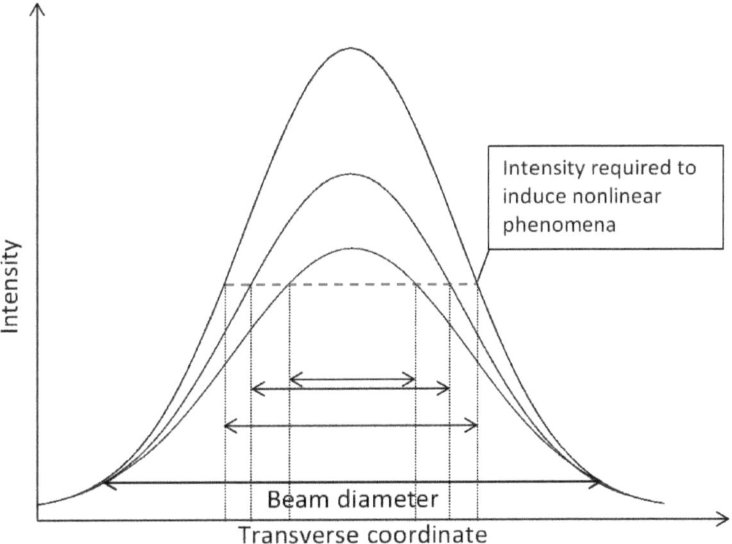

Figure 2. Focused laser beam spot diameter required to induce nonlinear effects dependency on radiation intensity.

Light-glass interactions used in DLW can induce different types of modifications. These depend on laser parameters such as radiation intensity, repetition rate, pulse duration, or wavelength to name a few [40]. Of course, induced modifications depend on the material itself [40,41]. Formally, in literature, it is possible to find 3 primary stable modification types that can be induced in the glass. Nevertheless, some exceptions apply at specific cases. For instance, silver clusters can be formed in silver containing glasses, resulting in type Argentum (or type A) modification [42]. However, in this review we will concentrate at the 3 primary modification types referring to any exceptions when it applies.

Type I modification is smooth material refraction index change. The change depends on various laser parameters and on the material itself. In a standard case, it is considered that such variation is mainly decided by pulse energy. An example photo of this modification can be seen in Figure 3b,c [40]. In the standard glass, the amplitude of change is in the order of 10^{-3}.

Type II of modification is periodical nanogratings in the glass. From optical standpoint, alongside changing just refractive index of the material, it also introduced birefringance. An example of nanogratings is shown in Figure 4. One of the key properties of this interaction is the tendency of nanogratings to be strongly influenced by the polarization of the incident beam. Nanograting ripple direction is perpendicular to polarization [41,43]. Moreover, the exact shape and properties of nanograting depend on other parameters as well. For instance, pulse repetition rate plays quite an important role, as it denotes the thermal aspect of the interaction [44]. The size of the ripples themselves is sub-diffraction, that is, in the range of tens of nm.

Type III modification denotes the formation of nano/micro-voids. When radiation intensity exceeds a particular value higher than needed for type II modification, micro-explosion in the volume of glass occurs and forms micro-cavities. This phenomenon is demonstrated in various works [45,46]. It is the most violent form of processing with an fs laser, leading to a relatively large modification area with several phase-change along the modified area. Particular radiation intensities to induce these types of modifications depend on the materials themselves.

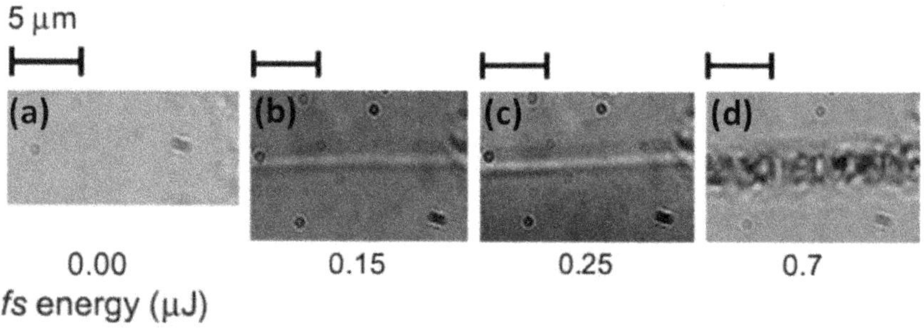

Figure 3. Glass modification examples. (**a**) unmodified material, (**b**,**c**) type I of glass modification or refractive index changes. (**d**)—type III modification or nano/micro-voids [40].

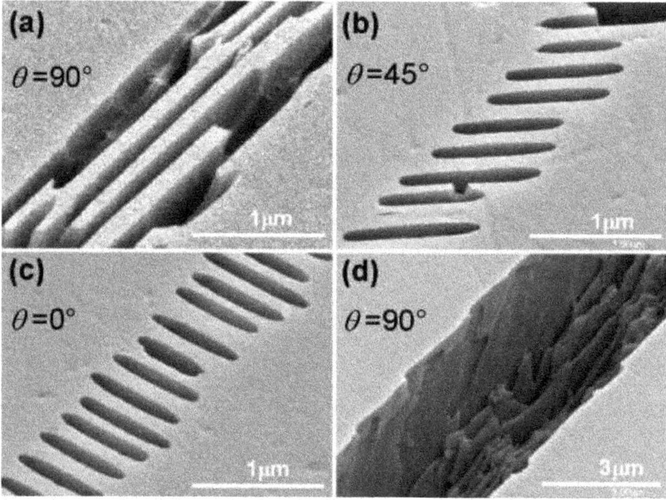

Figure 4. Examples of type II glass modification and visualization of nanogratings direction dependency to light polarization. (**a**) Nanogratings induced with light polarization perpendicular to scanning direction- the angle between polarization and scanning direction 90°, (**b**) 45° and (**c**) 0°. (**d**) Nanogratings modification after etching [41].

4. Fabrication of Functional 3d Structures

There are many laser glass processing techniques based on nonlinear light-matter interactions, and by inducing different types of modification it is possible to change the method by which we perform the processing of glass. Type I modification can be used for for smooth laser refractive index modification [40] and waveguides formation in glass [47,48], type II can be used for Q-plate manufacturing, while post-processing etching step of modification is necessary for selective laser etching (SLE) [49]. Meanwhile, type III modification is used for laser ablation [35] and welding [50]. These techniques have different natures and provide various possibilities of glass modification. We will describe various laser material modification techniques in more detail later.

4.1. Photonic Element Manufacturing

In a broad sense, the term photonics is used to encompass all advanced light-based science areas. In the light of this section, we will use it to describe very precise structures which exploit nm-level features and/or change in refractive index to control and/or confine light in accordance to application dictated requirements [51]. In general, manipulation of thin layers of materials with different kinds of refractive index is a very widely used phenomenon. One of the primary examples of it is dielectric mirrors [52]. The Bragg law for 2D system showing where the constructive interference is the strongest reads as:

$$2d \sin\theta = n_i \lambda, \tag{4}$$

where d is the distance between layers, θ—angle of incidence, λ—wavelength and n_i—positive integer. The most important parameter which can be controlled is the size of the features (d). Due to the already mention nonlinear nature of fs interaction with the transparent medium, it can be brought down to sub-diffraction level. Thus, one of the simplest examples of fs laser usage for photonic structure fabrication is the inscription of custom Bragg gratings into fibers [53]. In the simplest case, such structures are not strictly 3D. However, due to the flexibility of the technology fs laser allow forming fiber Bragg gratings (FBG) in arbitrary types of fiber, including non-photosensitized ones [54] or fibers made out of crystals [55]. At the same time, 3D capabilities are reflected by the capability to inscribe modifications in arbitrary place of the fiber, that is, not only in the core [56], or choose a specific core in multi-core fibers [57]. One of the main challenges here is focusing into the designated position within the fiber. Because the fiber shape normally is round, it defocuses the laser beam. Therefore, immersion oil [57], or flat ferules [58] might be used. This brings a lot of possibilities to the field, as it enables simple, on-demand fabrication of specialty FBGs which might be applied in various uses [53].

While FBG is a great example of a relatively simple photonic structure, 3D interactions could give substantially more diverse capabilities [51]. Therefore, the realization of 3D photonic crystals is quite an active field. Additive manufacturing was shown to be quite an attractive candidate [59]. However, in that case, the refractive index gradient between air and the polymer is around $\sim$0.5 which can be too sharp for some usages. At the same time, additively made 3D structures present objects on top of flat substrates, making them fragile and incompatible with anti-reflective coatings. In contrast, photonic elements embedded inside the volume of glass are robust, have a refractive index gradient of 10^{-2}–10^{-3} and could be potentially coated with the anti-reflective layer. Thus, glass-based 3D photonic devices occupy a very important niche in the field.

The first application is refractive index change in the glass is embedded 3D photonic crystals. It was shown that such photonic crystals can be used for spatial light filtering [60]. Indeed, the required spatial resolution and periodicity for such devices is close to the wavelength (in most cases HeNe 633 nm for simple testing) itself, making focused ultra-fast radiation a go-to tool. Simple woodpile-like geometries were used first. These were produced using both laser-assisted wet etching [61] and direct refractive index modification [62]. Nevertheless, these only allow filtering along X and Y axes of the photonic crystal, limiting applicability. A solution would be to use stacked ring geometry, which then would grant this effect along all 360° in a perpendicular direction to the propagation [63]. Here possibility to integrate such a device into glass volume becomes an enabling factor. If such a device would be tried using additive manufacturing additional supports would be needed to support each ring. Furthermore, standard single-period photonic crystals only allow limiting angular filtering. As a solution to expand it asymmetrically and chirped [62] geometry was proposed and tested, again attesting to the design flexibility enabled of 3D fs fabrication. The device was then successfully tested with microchip laser filtering in transverse modes (Figure 5), enabling it to increase its brightness by 2.8–3.1 times and single transverse mode from 88 mW to 335 mW [64]. If a one-directional filter is acceptable photonic crystal writing process can be extremely sped-up by applying Bessel beams [65]. In other spatial filtering techniques, the embedded photonic crystal is one of the most compact and simple solutions available, showing clear potential wide-spread applications.

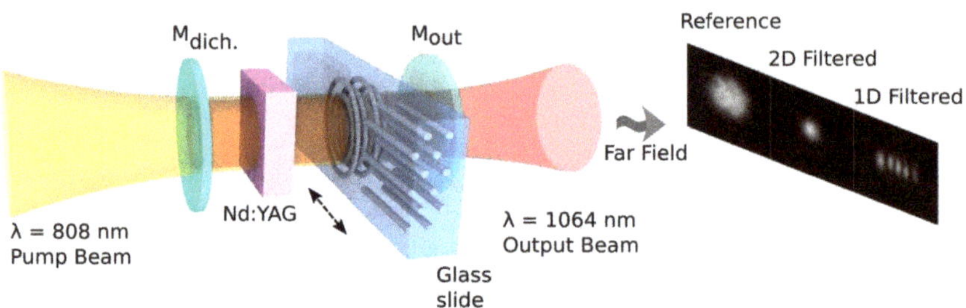

Figure 5. Principle of transverse mode filtering in microchip laser by application of laser-made photonic crystal embedded in a glass chip. The solution is compact and allows to achieve a high degree of control over the spatial characteristics of a laser, depending on the geometry of inserted element. Adopted from [64].

Laser-induced n modification can be exploited to not only shape light but also to confine and direct it. Indeed, embedded glass waveguides can be produced by using fs lasers [66]. Then, the structure has to meet the modification cross-section and Δn requirements. Interestingly, waveguides can be produced by both increasing and decreasing the n of the material. Regarding the direction, two primary modes are exploited. At first, focusing optics move up or down in relation to the surface of the sample. This is also called longitudinal writing geometry [67,68]. The advantage is a nearly perfectly round cross-section of the modification, as well as a very uniform gradient of n modification to all directions. Problems, on the other hand, lies in the fact that if the sample is thick, the working distance of sharply focusing optic might be insufficient. Thus, it is more suitable for lower NA optics. Also, if non-straight modifications are needed, the uniformity and high quality of the waveguide are compromised. Alternatively, the scanning might be performed alongside the horizontal plane, that is, in a transverse manner [69,70]. Then the length and shape of the waveguide can be arbitrary. On the other hand, due to Gaussian focusing properties, the modification is elongated in the Z direction, the Δn becomes highly non-uniform, especially in the Z direction. To remedy this additional elements, such as spatial light modulators (SLMs), might be needed in the optical system, complicating the setup and the process [71,72]. Alternatively, a multi-scan approach [73,74] or special light-matter interactions [75] might be exploited to make the cross-section of modification round. All of this can be combined with possibility to inscribe Bragg gratings into integrated waveguides, allowing creation of true functional 3D optical systems. It has to be stressed that it can be done in basically any kind of glass, including but not limited to borosilicate glass [76] or fused silica [77]. Thus, the needs of application dictate which of these two modes should be used. The waveguides themselves can be fabricated alone, or integrated into functional devices where they could act as a part of a sensor [78,79] or as an integrated interconnector between different types of devices [80]. This is especially attractive, as waveguide can potentially be integrated during the same technological step as the rest of the structure [78], owning to huge flexibility of fs-based processing of transparent mediums.

The elements discussed so far relied on type I modification. As a result, the designs and applicability of the produced devices were to some extent limited. To gain even greater tunability type II modification can be applied for free-form embedded optical element manufacturing. This type of fs modification allows inducing volume-embedded nanogratings which, due to orientation of sub-diffraction limited features, have distinct birefringence [81]. What is more, the directionality of the optical phenomena induced depends on the light polarization [82,83], giving huge controllability to the process. While the selection of materials in which this process is pronounced is not extremely broad [84–87], it is still sufficient to acquire a broad selection of possible optical devices. As a result it was employed to produce various optical structures, for instance diffraction gratings [88], Fresnel zone plates [89], computer-generated holograms [90,91], S-wave plates [92,93] or metamaterials [94]. Remarkable selection of possible optical elements is the result of both possibility to induce refractive index change and birefringence as well as the possibility to precisely control these modifications in 3D by using focused fs laser light. Additionally, these modifications exhibit immense thermal stability (up to 1000 °C) [95] allowing them to be considered for usage in some extreme environments. On top of that, as these elements are embedded into glass volume, their laser-induced damage threshold is also quite high ($\sim$J/cm^2) [96], that is, at the level of the host glass.

One extremely promising application of fs-laser-based embedded photonic structure fabrication is embedded optical memory. The idea to use light for encoding and reading memory dates back to the 1980s when the laser was employed to produce and read optical discs. Over the years due to progress in optical engineering, the wavelength used for the process could be reduced, in turn increasing the data density in the disk. This is due to the fundamental need to use diffraction-limited processes for both data recording and reading. This fundamentally limits achievable data density up to $\sim$0.25 Gb/cm^2 for standard systems using visible and near-UV systems [97]. However, as we discussed, fs-laser allows creating modifications not only on the surface of the material but also inside it paving the way for the volume memory [97,98]. Furthermore, single features created with fs laser can be extremely small (down to hundreds, or even tens of nm) allowing the extreme density of information [99]. Birefringence gives an additional degree of freedom as well, paving the way for true 5D memory (Figure 6) [100]. The modifications themselves have virtually unlimited lifetime [101], making them a valid candidate for extremely long-term big-data storage. Nevertheless, the modifications themselves can also be removed relatively simply if desired, granting the possibility to use the same medium to rewrite the information inside the same material volume. This can be achieved by subsequent laser exposure to the same volume [102]. Usage of crystals also allows to achieve it with relatively low temperatures (bellow 200 °C) [103] or exposure by loosely focused visible wavelength laser beam [104]. Admittedly, at the moment widespread adoption of this technology is limited due to the necessity to use expensive fs lasers and positioning. However, with the need to store up to massive amounts of information by numerous industries [105], this might become a more and more attractive solution for professional specialized extreme long-term data storage.

While the application of focused fs radiation has huge capabilities, there are some distinct limitations and hindrances native to light itself. First, if sharp focusing (NA > 0.4) is used, the polarization in the focal point has to be looked into more precisely than standard ray optics would suggest. Indeed, it is possible to show that sharply focusing fs light inside transparent medium changes E distribution in 3D [106,107]. Keeping in mind that the direction of type II modification heavily depends on the light polarization, this effect might have a severe influence on the result of fs processing and has to be accounted for. Furthermore, when working in the volume of a transparent medium, the laser beam is distorted due to aberrations [108]. This results in defusing effect, which changes the voxel's size and shape (aspect ratio) in different depths of the sample [109]. The functionality of photonic crystals and waveguides heavily depends on the shape of the modification as well as the precise modification of n. As a result, aberrations are substantial consideration in those applications. Luckily, this issue can be solved in several ways. First, focusing optics might have some tunability, allowing to account for different depths of fabrication [110,111]. The next step is the usage of SLM [109,112]. It was shown to be a powerful tool to account for focusing deficiencies and control the aspect ratio of the feature. For instance, 1:1 to 1:1.5 aspect ratio lines can be inscribed in any depth of 1 mm sample using 0.9 NA objective [109]. Alternatively, when SLM is not used, the aspect ratio can be more than 1:25. Additionally, due to defocusing, features could not be effectively written in depths below ~0.4 mm without significantly reoptimizing the process (i.e., increasing average laser power to compensate for defocusing. Even then, the feature aspect ratio would only increase with the depth. SLM also allows spatially shape incident beam (and, in turn, modification's shape in 3D) for advanced structure fabrication [72,113,114]. The process is flexible enough to account even for very high n, like diamond's 2.4 [115–117] or even anisotropic materials [118,119]. Unfortunately, SLM requires a very specialized setup, high precision in setup alignment, and advanced calculation algorithms alongside sufficient computing power [120]. On top of that standard SLMs are operating at relatively low frequencies of ~tens of kHz limiting the possibilities for dynamic possibilities of processing, with faster systems being investigated at the moment [121]. Finally, the LIDT of SLMs is relatively low, with high-end experimental devices capable of operating at P~100 W level using ultrafast lasers [122]. While solid-state SLMs might alleviate this issue [123], it remains a limiting factor. Thus, while SLMs offer a lot of interesting capabilities, their application remains limited.

Figure 6. An image depicting encoding of information using type II modification induced birefringence. While both of the images seem to be written in the same volume (image on the **left**) images of Maxwell (**center**) and Newton (**right**) can be separated by exploiting imaging along both optical axes. Taken from [100].

4.2. Selective Laser Etching (SLE)

Fundamentally, type II modification of glass leads to both, nanograting formation [43] and structural changes of material [124]. The photonic applications discussed so far mainly exploited the optical phenomena of inscribed modifications. Nevertheless, appreciation of chemical and mechanical changes can lead to a very interesting prospect for further applicability. Glass, or, more precisely, Si_2O compounds can be etched using hydrofluoric acid (HF) or Potassium Hydroxide (KOH). These etchants react with materials differently. The reaction equation of KOH and HF with one of the most popular glasses fused silica is shown in Equations (5) and (6). These reaction equations that concentrated KOH solution is beneficial over HF etchant due to least saturation behavior in elongating channel structures [125].

$$SiO_2 + 2KOH \rightarrow K_2SiO_3 + H_2O, \tag{5}$$

$$SiO_2 + 4HF \rightarrow SiF_4 + 2H_2O. \tag{6}$$

Normally, the etch rate of Fused silica is a few μm/h, of course, it depends on etchant [126]. Nevertheless, laser-induced type II modification can greatly expedite this process. Laser radiation induces material structural variations when binding angles of the lattice change. Oxygen atoms become more reactive and more effectively interacts with the etchant [124]. Also, formed nanogratings greatly increase the effective surface area at which etchant can interact with the material. Usually, for a description of this phenomenon definition of selectivity is used. Selectivity is the ratio of etching rates of modified and unmodified material. The utilization of this process allows obtaining selectivity up to 1000 [126]. Combined with the capability to inscribe such modifications in 3D it leads to the capability to selective glass etching (SLE) [124]. The most common material used in SLE is fused silica which is almost pure amorphous SiO_2. Fused silica SLE processing is a mostly studied process comparing to other materials. While other materials, such as borosilicate glass (BK7, Pyrex, Borofloat) [41,127,128], Foturan [129] can be used in SLE, these are rare occurrences. This is a result of lower selectivity [41] or requirement of post-processing [129,130], greatly favoring fused silica for widespread applications.

From the etchant side, HF and KOH are the most popular ones. KOH etching properties are strongly affected by KOH concentration and its temperature [131]. Usually, in the literature, we can find that 8–10 mol/L concentration KOH solution is used at 85–90 °C temperature [126,131,132]. Generally, a higher KOH concentration yield a higher etching rate. Nevertheless, a higher concentration of etchant does not guarantee better selectivity, potentially leading to contrary result [131]. This can be tied to the higher etching rate of both laser processed and unaffected areas. A similar trend can be observed with KOH temperature. When we are talking about the etching rate. It was shown, that the best selectivity is achieved at 80 °C [131]. Regarding the HF, if fast etching is desired, high (tens of %) HF concentrations should be used [133]. However, as one of the primary goals of SLE is achieving 3D structures, lower concentrations (5% or even lower) at ambient conditions are used to avoid overetching laser unexposed parts [126]. Curiously, experimentation on various conditions which might be used for SLE using HF (like temperature manipulation) is limited. Probably one of the main reasons, why more experiments are not performed which HF acid is a hazardous nature of this chemical. Subsequently, this is one of the reasons why more researcher groups prefer KOH over HF. Nevertheless, both etchants can find their applications. To get higher etching rates (up to a few hundred of μm/h) HF can be used [134]. Meanwhile, KOH should be used when high selectivity (up to 1000) is required [126], for instance producing high aspect ratio features. Of course, it is possible to combine both etchings one after another for one structure and use both etchant's advantages. This technique is called hybrid chemical etching [132,135].

One interesting prospect of SLE is the possibility to etch crystals like sapphire (Al_2O_3), Yttrium Aluminum Garnet ($Y_3Al_5O_{12}$ or YAG), lithium niobate ($LiNbO_3$) or crystalline quartz [136–139]. It is shown that nanogratings induced in a sapphire crystal are similar to ones inscribed to fused silica [140–142]. However, the etching mechanism of the crystals is slightly different from amorphous materials. When focused laser beam modifies crystalline sapphire modified region becomes amorphous [143]. Then, amorphous and porous regions are etched out in aggressive etchants like concentrated (40–50%) HF at room temperature [140–142,144–146] or around 35% KOH solution heated to 85–100 °C temperature [147]. Even more exotic etchant choices were demonstrated—sapphire was etched in phosphoric and sulfuric acid mixture at 300 °C temperature [148,149]. It showed that sapphire etched in 40% HF solution has selectivity which is in order of a few thousand (up to 5000) [140,141] and this value exceeds all the results showed on fused silica. YAG crystal shows even better selectivity which is estimated to be around 10^5 in HF acid. However, the SLE of crystals faces other problems: it is very hard to avoid cracks. Cracks are caused by modified crystalline material becoming amorphous and creating great volume tensions which lead to the crack of material [150]. Moreover, high selectivity comes from very low etching rates of unmodified material. Even modified material etching rate is very low, as little as up to 100 μm/h [137,140], for practical usage when a quite large structure is required such a low etching rate can be unacceptable.

Various laser parameters strongly affect the SLE process. A different aspect of selectivity, etched structure quality, such as surface roughness, can be decided by multiple laser parameters. Selectivity itself strongly depends on pulse energy. While sufficient pulse energy is needed to induce type II modification, the formation of LAZ can also be induced if too powerful pulses are used. LAZ affects etching properties [31]. Higher laser pulse energy with wider LAZ zones results in the wider etched zone. This effect could be for scanning up large volumes and, on contrary, this phenomena could be disadvantageous in high aspect ratio structure fabrication. Moreover, pulse duration and laser repetition rate affect LAZ and changes etching properties as well [44,151,152]. Longer pulse duration leads to stronger thermal effects. Higher pulse repetition rates prevent energy from being relaxed from the lattice. On the other hand, pulse energy not only change LAZ but also changes the size of modification and nanograting configuration itself, which leads to different selectivity induced by various pulse energies [41,43,151]. However, almost all this research leads to optimal particular parameter value which varies by changing other parameters. We can see that not individual parameters, but rather the whole parameters set is important for the SLE process. Basically, by changing pulse energy, pulse repetition rate the deposited dose is changed. Hence, a particular radiation dose is hiding under these parameter sets. Overall, it can be seen that for SLE process need to be ensured specific conditions—specific radiation dose [151] and these conditions can be altered by different thermal regime [31].

Alongside radiation parameters, scanning strategies also play an immense role in SLE effectiveness. Overall, translation velocity and pulse overlap while hatching/slicing denote the pulse overlap, which, in turn, governs accumulated radiation dose[151]. Therefore, one can consider that scanning velocity gives quite a similar effect as pulse repetition rate. As a result, selectivity dependency on scanning velocity was investigated in many works [131,151,153]. Acquired results varied, painting a more complicated picture of interaction as different sets of laser and scanning parameters were shown to work quite effectively [131,151]. Interestingly, in some works, almost no significant selectivity dependency on scanning speed was noticed [153]. The translation velocity question, in general, is very important, as it is one of the main parameters determining structuring rate, that is, what volume will be processed in a given amount of time thus better understanding of these contradicting results is a topic for further research in the field. Nevertheless, the scanning strategy also plays an important role. During the inscribing process, the specific separation between scanning lines should be chosen. Unsurprisingly, it also influences etching rate and selectivity [126,131]. One of the non-trivial observations made was that optimal Z separation value varies when different etchant is chosen [126]. Separation along a horizontal axis (i.e., slicing step) should also not exceed the voxel height [131]. Overall, due to the complex nature of the process and a huge amount of independent parameters, the exact interplay is still an object to an active investigation, with the goals of different groups distributed between trying to achieve the best selectivity, highest etching rate, lowest surface roughness or some combination of all.

It is important to not overlook the importance of polarization during the SLE process. Polarization is responsible for the direction of the nanogratings as depicted in Figure 4. Numerous works investigated selectivity dependency to laser beam polarization. When radiation polarization is perpendicular to scanning direction we get the highest etching rate [41,43]. The same applies and to selectivity because the unmodified material etching rate in all the cases remains the same [126]. Meanwhile polarization parallel to scanning direction gives the lowest etching rate [41,43]. Interestingly, circular polarization gives a good etching rate as in perpendicular polarization case [41]. This is extremely important because then the polarization might be considered invariable for any scanning strategy, heavily simplifying an already complicated process, becoming favorable for application-oriented fabrication. Nevertheless, care should be taken when considering the interplay between pulse duration and polarization. Discussed trend always appears when femtosecond laser pulses are used to write modification. Nevertheless, when ps pulses are used for SLE selectivity does not depend on radiation polarization [152]. This property comes from the nature of this process. By using femtosecond pulses it is possible to induce nanogratings modification, meanwhile, ps pulses yield modifications more reminiscent of nanocracks. These can be considered a different type of modification that is still suitable for SLE.

SLE stands in a relatively interesting place when considering piratical implementation. The process itself needs a lot of considerations in terms of laser parameters and writing strategies. On the other hand, it can be considered one of the most straightforward ways to produce 3D glass microstructures. It, as a result, possible applications were shown in microfluidics, micromechanics, optics, and photonics. In photonics, It is shown that SLE can be used for very precise photonic components fabrication. By using tightly focused laser beams it is possible to form very small (down to $\sim$100 nm) repeatable structures which after etching can be used as photonic components [137,148,149]. Micro optic element fabrication was also shown with SLE [130]. However, due to relatively high surface roughness after the SLE process ($\sim$hundred nm RMS) direct fabrication of optical components is impossible. Thus, after etching some annealing or other smoothing procedure is needed.

SLE excels in the field of microfluidics for the fabrication of lab-on-chip (LOC) devices. Glass is a preferred material for a lot of set applications, as it is chemically inert in organic solvents, transparent, and mechanically robust. SLE provides a relatively easy and highly controllable method for producing both surface and embedded channels. Here, SLE allows to form even complicated, curved, embedded 3D structures in glass [132,137,154,155]. It is demonstrated such channels ability to focused particles in the loops of the channels [137]. Other mentioned systems have a potential to be used in liquids filtering such as on-chip flow cytometry [137]. In the case of LOC microfluidic channel and some sort of integrated device inside it is formed [153,156]. Moreover, complicated microfluidic chains including filter and free movable integrated micro-switcher can be fabricated [132]. The mentioned device can perform fluid filtering. In the bigger channels narrower channels volume are inserted, only the smallest particles can go through the most narrow channels. Moreover, fluid flow direction can be controlled by fabricating channels with a free movable valve. Other 3D microfluidic devices which would be extremely challenging for any other technique can be fabricated [154]. To highlight some examples, Figure 7a–c show nozzle for biological applications. Figure 7d demonstrates connector for capillary electrophoresis. Figure 8 shows a microfluidic chip for cell sorting.

SLE also demonstrated huge promise in the field of micromechanics. For instance, a glass mechanical gripper can be produced and coated with metal to thermally induce movement [157,158]. A similar optical sensor can also be made [48]. The SLE-made mechanical device is coated by a metallic coat layer, which then moves when the potential changes. Information is red through the already integrated waveguide and the signal strength depends on the displacement of the structure. The beauty of this device is that it can be fabricated out of one piece of fused silica and at once it is possible to integrate waveguide and fabricate structure combining both type I and type II modifications in one fabrication step just by changing the laser parameters. Also, the device can be of arbitrary size, downscaled, and used as a component in more complex systems. Free movable micromechanical structures are also doable with SLE, such as gear systems [159]. Devices mentioned in the paragraph before conceptually are completely different from this micro-mechanical structure. In the first case, we get movement by fabricating structure with very high aspect ratio detail, so very thin details of glass become flexible. In the second case, a movable glass structure is fabricated because a slight gap is etched out between rotating and not rotating parts. Thus, by using the SLE technique, various types of movable structures can be obtained.

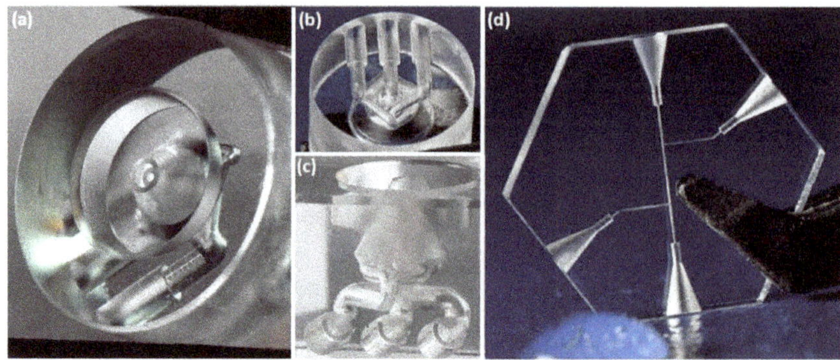

Figure 7. Example of microfluidic devices. (**a–c**) nozzle for biological applications, (**d**) glass connector for capillary electrophoresis, diameter 15 mm, thickness 2 mm [154].

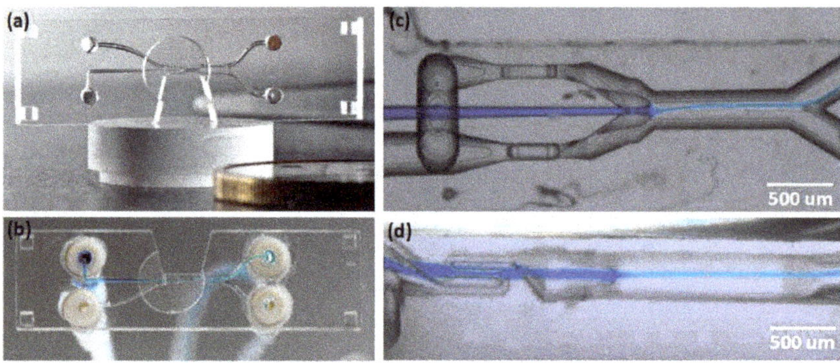

Figure 8. Example of microfluidic devices. (**a**,**b**) microfluidic chip created for cell sorting, (**c**,**d**) most important nodes of a chip are shown more detailed [154].

Finally, one must not forget that SLE is not an isolated tool and it can be combined with other DLW techniques, including multiphoton polymerization. Such fabrication is called hybrid fs laser manufacturing. For example, chemical liquid sensing application can be realized by combining SLE and multi-photon polymerization techniques [160]. Then, the channel and integrated cantilever are made out of glass in a single fabrication step. After the SLE process, a polymeric rod is produced by using multi-photon polymerization. When the whole structure is submerged into an organic solvent glass remains inert, while the rod expands or swells, a property well documented with laser-made 3D microstructures. Thus, such a device can act as a passive actuator or a mechanical sensor. This device is shown in Figure 9. Alternatively, it demonstrates that some hybrid devices can be used for cells sorting and detecting [161,162]. In closed SLE, the channels made are integrated with polymeric lend, which allows the detection of when the cell is moving through it as well as disturbing the signal.

Overall, SLE stands in quite a peculiar place in regards to its realization and applicability. The complexity of the process means realizing it is a rather complicated task, resulting in a steep learning curve for groups that attempt to do it. For this reason, a lot of researchers tend to use the substantially simpler type III modification base ablation, which will be discussed in the next subsection. Then, to achieve the same complex structures more fabrication steps of the simpler process might be needed. It also allows avoiding potentially hazardous HF-based chemical procedures. Alternatively, advanced high precision additive manufacturing-based solutions can be employed [163–165], forgoing the single step SLE process in favor of several simpler fabrication procedures. However, when SLE is employed right and all the technical nuances are taken care of, it provides unmatched design freedom and flexibility in producing arbitrary shaped 3D glass structures with resolution down to µm with overall size in the cm range.

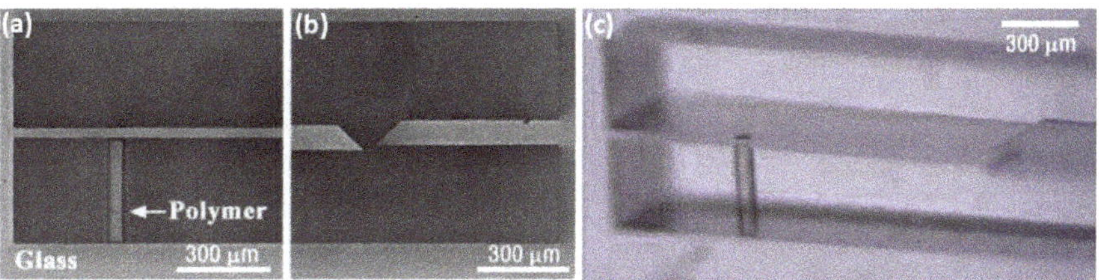

Figure 9. (**a–c**) Example of a hybrid microfluidic device [160].

4.3. Free-Form Cutting of Glasses

Fundamentally, direct laser ablation is the simplest of all DLW kinds. As the material is removed from the sample without any additional technological/post-processing steps it was quickly adopted in the industry using CW or long-pulsed (ns and longer) lasers [166]. Solid-state fs lasers are gaining popularity there, as well. The draw of ultrafast lasers in laser cutting is the possibility to control the thermal aspect of the process, subsequently allowing a smaller laser affected zone and cleaner cuts [23–25]. This capability is further expanded by the application of burst lasers, which allow both high structuring precision as well as very high throughput. There are also virtually no limitations for the application materials, which can be organic [167,168] (including living tissue [169]) dielectric [170–172] or metal [173,174]. However, when considering true 3D fabrication ablation have some severe drawbacks. In comparison to 3D printing [38], complex 3D architectures, arches, and integration of structures inside structures are either difficult or, for some geometries, impossible with direct ablation. Advanced techniques allow to achieve quasi-3D structures by either filliping [174] or turning samples [175] during processing, but it requires additional fabrication steps or a more complex setup, minimizing the main advantage of ablation—simplicity. Thus, in the following discussion, we will use the term 3D rather loosely, in many cases referring to structures that can otherwise be classified as 2.5D.

Among many applications where fs ablation of glasses can be exploited, microfluidics is one of the most common. Two types of channels can be produced in that case: on the glass surface [176–178], and embedded inside the volume [179,180]. Surface channels are substantially simpler to produce. Also, they allow integration of additional elements using other fabrication techniques, for instance, multiphoton polymerization (MPP) [181]. However, in that case, an additional channel sealing step is needed. It can be realized in numerous ways, including direct laser welding. It will be discussed in more detail further in the article. Nevertheless, it is an additional technological step that, if possible, should be avoided. Alternatively, channels inside the bulk of the material can be produced. Indeed, fs lasers allow surprisingly very fine control of type III modifications in glass volume [182]. Despite this, if ablation is used there is no direct way for debris to leave the embedded channel if this method is used. Luckily, if the process is carried out in the water, the debris removal is expedited, resulting in the possibility to produce clean and complex channel systems [179,180], or high aspect ratio holes (for instance 4 µm wide and more than 200 µm deep [183]). By applying Bessel beam ablation it is shown even higher aspect ratio narrow channels which is around 200 nm in diameter and 20 µm in depth [184]. Alternatively, water can assist in the structuring of the volume of porous/photosensitive glass [185–188]. Then, after laser exposure, glass is heated up and solidified, embedding produced 3D system inside a material. Additional etching and annealing steps can also follow. However, then the process becomes somewhat similar in complexity (if not more difficult to perform) to SLE yielding substantially worse shape control and surface finish of the channel walls.

At the same time, non-SLE methods allow using potentially simpler glasses. Regardless, apart from several demonstrations, this methodology is used sparingly.

Microoptics can also be produced using direct laser ablation. While extremely complex designs achievable using additive manufacturing are out of the question [189–191], ablation exceeds those methods by allowing to use of glass and achieve overall size from tens of μm to mm. Optical elements, like various type lenses [192–194] or axicons [195], can be produced this way. In virtually all the cases additional post-processing steps are mandatory due to the necessity to smooth the surface of the structure down to optical quality (less than $\lambda/10$). Annealing is the most popular option. It gently melts the surface of the optical element, allowing surface tension to increase surface smoothness without compromising the overall shape (Figure 10). This can be induced by heating the structure to a sufficient temperature (close to the melting of the material) [196], or by employing another laser such as CO_2 [192–195]. An alternative way to achieve lenses using type III laser modification is based on the idea, of inducing very fine modifications on the surface of the material. Then the sample is submerged to the etchant, which starts to etch out the material. The modification acts as a seed, which expands over time. Due to the nature of the process, expanded areas form lens-like profiles and, after some time reach each other [197,198]. Thus, a plano concave lens array is formed with an optical grade surface finish. Overall, while ablation allows achieving free-form optics, etching provides superior shape control and a possibility to produce more complex geometries, making both approaches somewhat comparable.

Surface roughness after type III processing is rather poor ($\sim$several μm RMS). A laser can be employed to smooth it out. Normally, CO_2 [199] or high power solid-state lasers [200] are employed for this task. The process is based on inducing desired thermal interactions at the surface of the sample. It can be used to polish glass parts produced using SLE [201] or ablation [202]. However, the fs laser itself can be used for this task. The primary advantage of using ultrashort pulses is the possibility to extremely precisely control thermal accumulation on the surface of the sample [203]. This can be exploited to polish some exotic materials, like Germanium [204] or yttria-stabilized zirconia [205]. Additionally, due to the controlled manner of the process fs lasers allow avoiding some of the effects associated with polishing using different laser sources or mechanical polishing. Nevertheless, the application of fs lasers in polishing remains sparse. At the same time, the advent of burst-mode capable systems brings a new level of control of thermal aspects of the interaction, potentially reducing surface roughness during processing in the first place, bringing new capabilities to the field [25].

All the processes showcased so far mostly rely on Gaussian beam focusing using standard optics. However, as discussed previously, light is a flexible tool that can be spatiotemporally manipulated. To change the spatial distribution of the laser beam, passive elements can be inserted into the path of the laser beam. In more advanced cases, it can be done using SLM, which allows the changing of a projected mask dynamically on-demand [206]. However, due to the relatively low LIDT of such elements, their usage in direct ablation might be limited. Regarding the exact applications various examples exist. For instance, the Bessel beam can allow taper-less fabrication [184]. In other words, the channel walls are nearly vertical, which is hard to achieve with sharply focused Gaussian beams. It also can be used to achieve high aspect ratio (up to 1:100) features easily [207]. Also, in the case of top-hat laser beam distribution, the cut quality can be improved [208]. Beam shaping also can be exploited to create entire patterns to be ablated [206]. Regarding the temporal properties of fs radiation, it can also be tuned to achieve desirable ablation properties [209]. Manipulation of environmental aspects should also not be overlooked. By changing the pressure during the ablation process the debris formation tendencies can be influenced, which helps with the overall quality of the sample [210]. Alternatively, additional gasses are shown to help increase ablation depth [211]. Water, which was shown to be a rather popular additive in fs glass machining, can also be used to induce spontaneous spatiotemporal light filamentation. It can be used to cut arbitrary materials, including glass, at relatively large thicknesses (1 mm and more) [212]. Thus, overall, while base fs laser ablation of glasses might be considered somewhat limited as far as 3D fabrication is concerned, some capabilities exist to enhance the result of processing. At the same time, then it somewhat complicated the processing, making it less attractive.

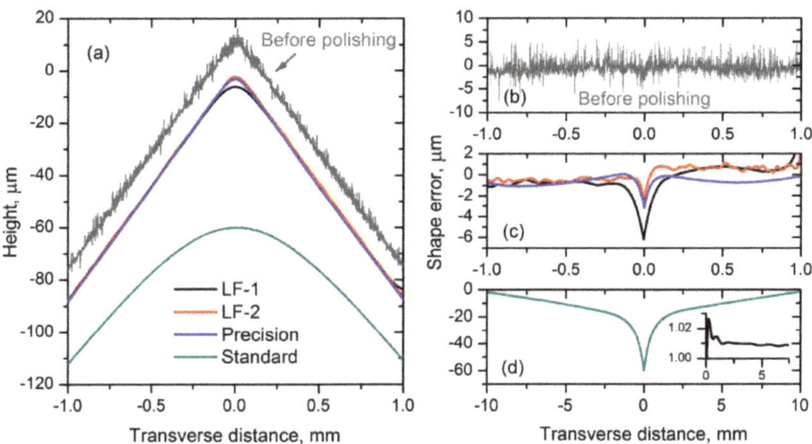

Figure 10. The precision of profile of the ultrafast laser-made axicon with subsequent CO_2 polishing and comparison to commercial high-precision and standard counterparts. The un-polished and polished laser-made profiles are offset to better reveal the difference between them. Reproduced from [195].

4.4. Laser Welding

Discussing 3D structure fabrication it is important to understand that in most cases a 3D structure is, in fact, multiple stacked 2D layers. Only a few processing techniques, for instance, selective laser etching, allow true 3D fabrication, yet it also relies hard on layering as it allows to minimize effects like shadowing. Sometimes the stacking is part of the process itself, like in 3D printing, while sometimes additional bonding step is required. Except for integrated free-form SLE structures, stacking glass onto glass to form 3D structures is also common, especially when parts are produced using ablation. There exist different bonding techniques such as direct, adhesive bonding, or others. Adhesive bonding is performed by using additional intermediate chemical materials [213], but it has limitations to strength as well as adds additional chemical constitutes which might be detrimental to some application. Direct bonding is performed by increasing the temperature of the samples, it starts to melt and connects with each other [214]. Thus, direct bonding by ultrafast laser welding becomes an attractive way to bond glass layers with each other forgoing most of the listed problems. Laser welding relies on type III modifications [215,216]. It can be considered one of the most advanced available bonding techniques, because of high spatial selectivity, localized material modification, and capability of completely sealing functional devices. Moreover, laser welding allows to achieve up to 95% breaking resistance of primary material [217,218]. What is more, laser welding does not require additional materials or other post-processing to be included. It is shown that laser welding can be used as a supplementary technology to strengthen other bonding techniques [219].

To weld two samples laser beam needs to be focused on the contact of those samples. Because of high radiation intensity, tunneling or multi-photon ionization occurs. Radiation is absorbed only in the focal spot where free-electron appears which mount increase because of avalanche ionization—plasma forms. Meanwhile, surrounding materials heats which enhance nonlinear which then, in turn, expedites the heating. The volume of melted material encompasses both samples which need to be bonded. Afterward, the material cools and forms a firm connection between samples [220]. By melting material some of its chemical connections are broken. In the affected material some oxygen connection remains open forming a new chemical connection between samples [221]. After measuring welded samples cross-section Raman spectra was found that welded different material samples mix in the contact [222]. Thus, this explains high welded material breaking resistance.

For a process like this optical contact is required. Optical contact appears because of inter-molecular Van der Vals bonds. If the right conditions are created for two materials surfaces to be close enough to each other, those samples start to attract each other because of mentioned forces (forces decrease proportionally to the square of the distance). Thus, in this way, we get stable optical contact [223], which does not require to be additionally supported. Even more, additional external forces applied to the samples leads to cracks in the welded seam [224]. Therefore, clean and high-quality surfaces practically became a requirement for the laser welding process. Spacing between two samples should not be larger than a few hundreds of nm, otherwise, it is hard to avoid cracks in the seam or even ablation of one of the surfaces [225]. Optical contact is one of the greatest challenges for laser welding, because often it is hard to achieve it due to surfaces imperfection of the samples, due to high surface area, or imperfections/undulations in the sample caused by other processing techniques, like ablation.

Advances in laser engineering positively influence the ease with which welding can be realized. It was shown how to avoid the requirement of optical contact. Optical contact is still preferred, allowing us to reach a higher breaking resistance of a welded seam than welding without optical contact can [50]. However, for simpler, less challenging applications even simplified welding is sufficient. At this moment, it ihas been shown that laser welding of samples with gaps between each other up to 3 μm [220]. There a few factor with allows to walk around this requirement. First of all, the gaps which could be efficiently filled depends on the amount of material that is brought to the gap. This amount of material can be evaluated as the size of a welding seam. Seam size directly depends on laser radiation and scanning parameters (pulse energy, pulse repetition rate or scanning velocity, focus position in the Z direction in respect of the contact of samples) used for the welding process or even it could depend on material thermal conductivity. Thus, it possible to lower the high-quality contact requirement by increasing the size of the seam. Welded seam size can be increased in a few ways. The usage of high repetition rate (hundreds of kHz) radiation leads to a greater size of a seam [225]. The same applies to higher pulse energies [50,218,221]. This dependency is schematically shown in Figure 11b. By bringing more materials to the contact of the sample, it not only provides the possibility to weld samples with larger gaps but also affects the breaking resistance of welded samples. A larger welding seam will lead to higher breaking resistance [217]. Care should be taken, however, because of the possibilities of breaking material due to high internal tension instead of material welding [226,227].

It is important to understand how scanning properties affect the size of the seam. Lower scanning velocity will lead to higher pulse overlapping resulting in more energy being delivered to each point. As a consequence larger welding seam and higher breaking resistance are obtained [221,228]. What is more, the gap between separate seams, or hatching step, affects samples breaking resistance significantly [50,218]. It is quite intuitive that a larger amount of seams in the same sample allow the obtaining of a greater breaking resistance. On the other hand, welding sample by separate seams brings its own benefits to this process. Distinguished seams give a possibility to contact materials with different thermal expanding coefficients. Because the connection is very local, the spacing between seam compensates for the fact that the two materials expand differently. Unsurprisingly, the focal spot position in the Z direction in respect of gap position is also a critical parameter in laser welding. The way this parameter influences the process is shown in Figure 11a. It can be observed that, in laser welding, for the most effective process the focus of the radiation should be a little bit ($\sim$100 μm) below the contact of the samples [220]. In this way, melted material comes up to the contact and fills it in the most efficient way. This tactic allows us to melt more material and fill the gap between samples with the greatest amount of material.

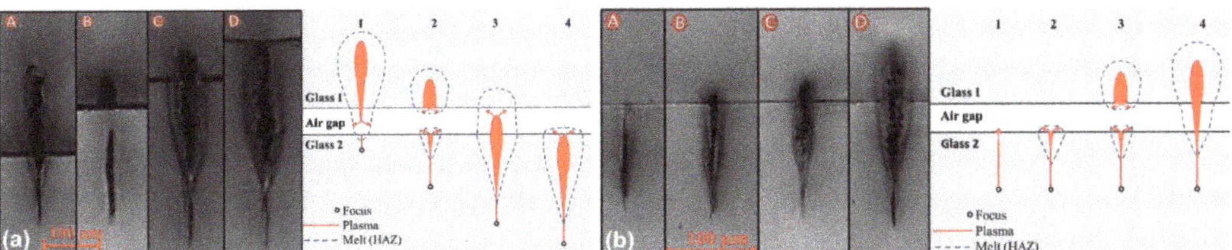

Figure 11. Cross-section of the welding process (**a**) welding dependency on focal position. The focal position is changed in this order from image A to D: −24.1 µm, −40.0 µm, −86.5 µm and −107.3 µm, Corresponding schemes of these process are shown from picture 1 to 4, (**b**) welding dependency on pulse energy, the pulse energy is increasing in this order from image A to D: 10.1, 11.23, 12.9 and 18.8 µJ, schemes of the corresponding process are shown in pictures 1 to 4 [220].

The advent of burst lasers have also been experimented with in welding. It allows improvement of the welding process in many aspects. Burst-welds have higher breaking resistance than that achieved by only welding in a high repetition rate regime [229]. Moreover, usage of burst enables us to weld materials with greater spacing in the contact between samples [50]. Another interesting approach is laser welding with Bessel beams [222]. One of the unique Bessel beams property is long focal length. Its relay length can up to a few hundred µm. Thus, in this way we do not need to have accurate information about the position of the contact of materials.

Laser welding can be applied in many functional device fabrication. It can be applied in such areas as electronics [230], microfluidics [213]. This technology could be extremely useful in electronics when the whole chip cannot be processed in other bonding techniques and local laser welding means that important parts of the chip can be left unaffected. Ultrafast welding is also advantageous, as it allows to easily bond glasses with other materials like metals or polymers. However, due to high technical requirements for both laser systems and surfaces of bonded samples, the advance of this technique is so far slow, with most demonstrations being done only in academia with proof-of-concept work.

5. Conclusions

The field fs laser-based processing of glasses and transparent dielectrics is wast. In this review, we have shown that:

1. Due to the nonlinear nature of fs pulses-based light-matter interaction, transparent to the used wavelength dielectrics can be processed, including glasses and crystals. This opens doors for true 3D processing, including inside the volume of the material. Due to interaction peculiarities, three distant types of modifications can be formed. Type I results in a localized smooth refractive index change in the range of $\Delta n = 10^{-3}$. This can be used to create embedded waveguides, Bragg gratings,and other integrated photonic structures.
2. Type II modification is created with higher pulse energy, which manifests as embedded, sub-diffraction-limited nanograting. Characteristics of such ripples depend on pulse energy and polarization. They also have pronounced birefringence. These can be used to produce various optical and photonic elements. If put into an etching solution, such as HF or KOH, these also etch out from tens to thousands of times faster, enabling selective 3D glass etching.
3. Due to the nature of type III modification, that is, ablation can be used for applications demanding less precision but higher precision, such as microfluidics. Application in optics is also possible, but additional post-processing is needed. Overall, this processing type partially sacrifices true 3D processing capability in exchange for being a lot simpler and faster.

4. As fs lasers allow us to precisely control thermal effects during processing, they can be employed for polishing and welding. In the latter case, it allows the joining of glasses of the same and different types, as well as glasses with plastics or metals, making it one of the most versatile tools available for such applications.

Overall, the key advantage of the fs laser is flexibility. Such processing can allow achievement of the same 3D structure using very different methodologies, allowing us to choose a process on a case-by-case basis. With this flexibility, one can expect the field of fs pulse-based glass processing to continue to grow rapidly in the near future, with both wider applications of existing methodologies and the creation of new ones.

Author Contributions: A.B. prepared sections of radiation and material interaction, selective laser etching and laser welding. L.J. prepared sections about photonic elements manufacturing and laser cutting. All authors have read and agreed to the published version of the manuscript.

Funding: This research was funded by EC Horizon 2020 program, ATOPLOT project, grant number 950785.

Acknowledgments: In this section you can acknowledge any support given which is not covered by the author contribution or funding sections. This may include administrative and technical support, or donations in kind (e.g., materials used for experiments).

Conflicts of Interest: The authors declare no conflict of interest.

References

1. Ivey, D.G.; Northwood, D.O. Storing energy in metal hydrides: A review of the physical metallurgy. *J. Mater. Sci.* **1983**, *18*, 321–347. [CrossRef]
2. Dewidar, M.M.; Yoon, H.C.; Lim, J.K. Mechanical properties of metals for biomedical applications using powder metallurgy process: A review. *Met. Mater. Int.* **2006**, *12*, 193–206. [CrossRef]
3. Thames, S.F.; Panjnani, K.G. Organosilane polymer chemistry: A review. *J. Inorg. Organomet. Polym.* **1996**, *6*, 59–94. [CrossRef]
4. Merkininkaitė, G.; Gailevičius, D.; Šakirzanovas, S.; Jonušauskas, L. Polymers for Regenerative Medicine Structures Made via Multiphoton 3D Lithography. *Int. J. Polym. Sci.* **2019**, *2019*, 1–23. [CrossRef]
5. Feng, X.; Qi, C.; Lin, F.; Hu, H. Tungsten–tellurite glass: A new candidate medium for Yb3-doping. *J. Non-Cryst. Solids* **1999**, *256–257*, 372–377. [CrossRef]
6. Colomban, P.; Schreiber, H.D. Raman signature modification induced by copper nanoparticles in silicate glass. *J. Raman Spectrosc.* **2005**, *36*, 884–890. [CrossRef]
7. Goh, Y.F.; Alshemary, A.Z.; Akram, M.; Kadir, M.R.A.; Hussain, R. Bioactive Glass: AnIn-VitroComparative Study of Doping with Nanoscale Copper and Silver Particles. *Int. J. Appl. Glass Sci.* **2014**, *5*, 255–266. [CrossRef]
8. Ahmadi, F.; Ebrahimpour, Z.; Asgari, A.; El-Mallawany, R. Role of silver/titania nanoparticles on optical features of Sm3+ doped sulfophosphate glass. *Opt. Mater.* **2020**, *105*, 109922. [CrossRef]
9. Inoue, A.; Shinohara, Y.; Yokoyama, Y.; Masumoto, T. Solidification Analyses of Bulky Zr60Al10Ni10Cu15Pd5 Glass Produced by Casting into Wedge-Shape Copper MoldSolidification Analyses of Bulky Zr60Al10Ni10Cu15Pd5 Glass Produced by Casting into Wedge-Shape Copper Mold. *Mater. Trans. JIM* **1995**, *36*, 1276–1281. [CrossRef]
10. Laws, K.J.; Gun, B.; Ferry, M. Influence of Casting Parameters on the Critical Casting Size of Bulk Metallic Glass. *Metall. Mater. Trans. A* **2009**, *40*, 2377–2387. [CrossRef]
11. Fujita, K.; Morishita, Y.; Nishiyama, N.; Kimura, H.; Inoue, A. Cutting Characteristics of Bulk Metallic Glass. *Mater. Trans.* **2005**, *46*, 2856–2863. [CrossRef]
12. Foy, K.; Wei, Z.; Matsumura, T.; Huang, Y. Effect of tilt angle on cutting regime transition in glass micromilling. *Int. J. Mach. Tools Manuf.* **2009**, *49*, 315–324. [CrossRef]
13. Huang, C.C.; Hewak, D.W.; Badding, J.V. Deposition and characterization of germanium sulphide glass planar waveguides. *Opt. Express* **2004**, *12*, 2501–2506. [CrossRef] [PubMed]
14. Ostomel, T.A.; Shi, Q.; Tsung, C.K.; Liang, H.; Stucky, G.D. Spherical Bioactive Glass with Enhanced Rates of Hydroxyapatite Deposition and Hemostatic Activity. *Small* **2006**, *2*, 1261–1265. [CrossRef]
15. Tay, F.E.H.; Iliescu, C.; Jing, J.; Miao, J. Defect-free wet etching through pyrex glass using Cr/Au mask. *Microsyst. Technol.* **2006**, *12*, 935–939. [CrossRef]
16. Iliescu, C.; Chen, B.; Miao, J. On the wet etching of Pyrex glass. *Sens. Actuators A-Phys.* **2008**, *143*, 154–161. [CrossRef]
17. Von Witzendorff, P.; Pohl, L.; Suttmann, O.; Heinrich, P.; Heinrich, A.; Zander, J.; Bragard, H.; Kaierle, S. Additive manufacturing of glass: CO_2-Laser glass deposition printing. *Proc. CIRP* **2018**, *74*, 272–275. [CrossRef]
18. Zaki, R.M.; Strutynski, C.; Kaser, S.; Bernard, D.; Hauss, G.; Faessel, M.; Sabatier, J.; Canioni, L.; Messaddeq, Y.; Danto, S.; et al. Direct 3D-printing of phosphate glass by fused deposition modeling. *Mater. Des.* **2020**, *194*, 108957. [CrossRef]

19. Kotz, F.; Arnold, K.; Bauer, W.; Schild, D.; Keller, N.; Sachsenheimer, K.; Nargang, T.M.; Richter, C.; Helmer, D.; Rapp, B.E. Three-dimensional printing of transparent fused silica glass. *Nature* **2017**, *544*, 337–339. [CrossRef]
20. Kotz, F.; Quick, A.S.; Risch, P.; Martin, T.; Hoose, T.; Thiel, M.; Helmer, D.; Rapp, B.E. Two-Photon Polymerization of Nanocomposites for the Fabrication of Transparent Fused Silica Glass Microstructures. *Adv. Mater.* **2021**, *33*, 2006341. [CrossRef]
21. Sibbett, W.; Lagatsky, A.A.; Brown, C.T.A. The development and application of femtosecond laser systems. *Opt. Express* **2012**, *20*, 6989–7001. [CrossRef]
22. Jonušauskas, L.; Mackevičiūtė, D.; Kontenis, G.; Purlys, V. Femtosecond lasers: The ultimate tool for high-precision 3D manufacturing. *Adv. Opt. Technol.* **2019**, *8*, 241–251. [CrossRef]
23. Chichkov, B.N.; Momma, C.; Nolte, S.; Alvensleben, F.; Tünnermann, A. Femtosecond, picosecond and nanosecond laser ablation of solids. *Appl. Phys. A* **1996**, *63*, 109–115. [CrossRef]
24. Leitz, K.H.; Redlingshöfer, B.; Reg, Y.; Otto, A.; Schmidt, M. Metal Ablation with Short and Ultrashort Laser Pulses. *Phys. Procedia* **2011**, *12*, 230–238. [CrossRef]
25. Kerse, C.; Kalaycıoğlu, H.; Elahi, P.; Çetin, B.; Kesim, D.K.; Akçaalan, Ö.; Yavaş, S.; Aşık, M.D.; Öktem, B.; Hoogland, H.; et al. Ablation-cooled material removal with ultrafast bursts of pulses. *Nature* **2016**, *537*, 84–88. [CrossRef]
26. Gattass, R.R.; Mazur, E. Femtosecond laser micromachining in transparent materials. *Nature Photon.* **2008**, *2*, 219–225. [CrossRef]
27. Brabec, T.; Krausz, F. Intense few-cycle laser fields: Frontiers of nonlinear optics. *Rev. Mod. Phys.* **2000**, *72*, 545–585. [CrossRef]
28. Keldysh, L.V. Ionization in the field of a strong electromagnetic wave. *J. Exp. Theor. Phys.* **1965**, *20*, 1307–1314.
29. Schaffer, C.B.; Brodeur, A.; Mazur, E. Laser-induced breakdown and damage in bulk transparent materials induced by tightly focused femtosecond laser pulses. *Meas. Sci. Technol.* **2001**, *12*, 1784–1794. [CrossRef]
30. Sundaram, S.K.; Mazur, E. Inducing and probing non-thermal transitions in semiconductors using femtosecond laser pulses. *Nat. Mater.* **2002**, *1*, 217–224. [CrossRef]
31. Bellouard, Y.; Said, A.; Dugan, M.; Bado, P. Fabrication of high-aspect ratio, micro-fluidic channels and tunnels using femtosecond laser pulses and chemical etching. *Opt. Express* **2004**, *12*, 2120–2129. [CrossRef] [PubMed]
32. Zhimalov, A.B.; Solinov, V.F.; Kondratenko, V.S.; Kaplina, T.V. Laser cutting of float glass during production. *Glass Ceram.* **2006**, *63*, 319–321. [CrossRef]
33. Zhao, C.; Zhang, H.; Wang, Y. Semiconductor laser asymmetry cutting glass with laser induced thermal-crack propagation. *Opt. Lasers Eng.* **2014**, *63*, 43–52. [CrossRef]
34. Kanemitsu, Y.; Tanaka, Y. Mechanism of crack formation in glass after high-power laser pulse irradiation. *J. Appl. Phys.* **1987**, *62*, 1208–1211. [CrossRef]
35. Mirza, I.; Bulgakova, N.M.; Tomaštik, J.; Michalek, V.; Haderka, O.; Fekete, L.; Mocek, T. Ultrashort pulse laser ablation of dielectrics: Thresholds, mechanisms, role of breakdown. *Sci. Rep.* **2016**, *6*, 1–11. [CrossRef]
36. Dyer, P.; Farley, R.J.; Giedl, R.; Karnakis, D. Excimer laser ablation of polymers and glasses for grating fabrication. *Appl. Surf. Sci.* **1996**, *96–98*, 537–549. [CrossRef]
37. Buerhop, C.; Blumenthal, B.; Weissmann, R. Glass surface treatment with excimer and CO_2 lasers. *Appl. Surf. Sci.* **1990**, *46*, 430–434. [CrossRef]
38. Jonušauskas, L.; Juodkazis, S.; Malinauskas, M. Optical 3D printing: Bridging the gaps in the mesoscale. *J. Opt.* **2018**, *20*, 053001. [CrossRef]
39. Jesacher, A.; Booth, M.J. Parallel direct laser writing in three dimensions with spatially dependent aberration correction. *Opt. Express* **2010**, *18*, 21090–21099. [CrossRef]
40. Krol, D. Femtosecond laser modification of glass. *J. Non-Cryst. Solids* **2008**, *354*, 416–424. [CrossRef]
41. Hnatovsky, C.; Taylor, R.S.; Simova, E.; Rajeev, P.P.; Rayner, D.M.; Bhardwaj, V.; Corkum, P.B. Fabrication of microchannels in glass using focused femtosecond laser radiationand selectivechemical etching. *Appl. Phys. A* **2006**, *84*, 47–61. [CrossRef]
42. Khalil, A.A.; Bérubé, J.P.; Danto, S.; Cardinal, T.; Petit, Y.; Canioni, L.; Vallée, R. Comparative study between the standard type I and the type A femtosecond laser induced refractive index change in silver containing glasses. *Opt. Mater. Express* **2019**, *9*, 2640–2651. [CrossRef]
43. Hnatovsky, C.; Taylor, R.S.; Simova, E.; Bhardwaj, V.R.; Rayner, D.M.; Corkum, P.B. Polarization-selective etching in femtosecond laser-assisted microfluidic channel fabrication in fused silica. *Opt. Lett.* **2005**, *30*, 1867–1869. [CrossRef]
44. QI, J.; Wang, Z.; Xu, J.; Lin, Z.; Li, X.; Chu, W.; Cheng, Y. Femtosecond laser induced selective etching in fused silica: Optimization of the inscription conditions with a high-repetition-rate laser source. *Opt. Express* **2018**, *26*, 29669–29678. [CrossRef] [PubMed]
45. Itoh, K.; Watanabe, W.; Nolte, S.; Schaffer, C.B. Ultrafast processes for bulk modification of transparent materials. *MRS Bull.* **2006**, *31*, 620–625. [CrossRef]
46. Sugioka, K.; Cheng, Y. Ultrafast lasers-reliable tools for advanced materials processing. *Light Sci. Appl.* **2014**, *3*, 4. [CrossRef]
47. Miura, K.; Qiu, J.; Inouye, H.; Mitsuyu, T.; Hirao, K. Photowritten optical waveguides in various glasses with ultrashort pulse laser. *Appl. Phys. Lett.* **1997**, *71*, 3329–3331. [CrossRef]
48. Bellouard, Y.; Said, A.A.; Bado, P. Integrating optics and micro-mechanics in a single substrate: A step toward monolithic integration in fused silica. *Opt. Express* **2005**, *13*, 6635–6644. [CrossRef]
49. Liu, X.Q.; Bai, B.F.; Chen, Q.D.; Sun, H.B. Etching-assisted femtosecond laser modification of hard materials. *Opto-Electron. Adv.* **2019**, *2*, 190021. [CrossRef]

50. Richter, S.; Zimmermann, F.; Eberhardt, R.; Tunnermann, A.; Nolte, S. Toward laser welding of glasses without optical contacting. *Appl. Phys. A* **2015**, *121*, 1–9. [CrossRef]
51. Li, W.; Meng, F.; Chen, Y.; Li, Y.; Huang, X. Topology Optimization of Photonic and Phononic Crystals and Metamaterials: A Review. *Adv. Theory Simul.* **2019**, *2*, 1900017. [CrossRef]
52. Apfel, J.H. Phase retardance of periodic multilayer mirrors. *Appl. Opt.* **1982**, *21*, 733. [CrossRef]
53. Pallarés-Aldeiturriaga, D.; Roldán-Varona, P.; Rodríguez-Cobo, L.; López-Higuera, J.M. Optical Fiber Sensors by Direct Laser Processing: A Review. *Sensors* **2020**, *20*, 6971. [CrossRef]
54. Wikszak, E.; Thomas, J.; Burghoff, J.; Ortaç, B.; Limpert, J.; Nolte, S.; Fuchs, U.; Tunnermann, A. Erbium fiber laser based on intracore femtosecond-written fiber Bragg grating. *Opt. Lett.* **2006**, *31*, 2390–2392. [CrossRef]
55. Grobnic, D.; Mihailov, S.; Smelser, C.; Ding, H. Sapphire Fiber Bragg Grating Sensor Made Using Femtosecond Laser Radiation for Ultrahigh Temperature Applications. *IEEE Photon. Technol. Lett.* **2004**, *16*, 2505–2507. [CrossRef]
56. Lai, Y.; Zhou, K.; Sugden, K.; Bennion, I. Point-by-point inscription of first-order fiber Bragg grating for C-band applications. *Opt. Express* **2007**, *15*, 18318–18325. [CrossRef] [PubMed]
57. Wolf, A.; Dostovalov, A.; Bronnikov, K.; Babin, S. Arrays of fiber Bragg gratings selectively inscribed in different cores of 7-core spun optical fiber by IR femtosecond laser pulses. *Opt. Express* **2019**, *27*, 13978–13990. [CrossRef] [PubMed]
58. Dostovalov, A.V.; Wolf, A.A.; Parygin, A.V.; Zyubin, V.E.; Babin, S.A. Femtosecond point-by-point inscription of Bragg gratings by drawing a coated fiber through ferrule. *Opt. Express* **2016**, *24*, 16232–16237. [CrossRef] [PubMed]
59. Maigyte, L.; Purlys, V.; Trull, J.; Peckus, M.; Cojocaru, C.; Gailevičius, D.; Malinauskas, M.; Staliunas, K. Flat lensing in the visible frequency range by woodpile photonic crystals. *Opt. Lett.* **2013**, *38*, 2376–2378. [CrossRef]
60. Maigyte, L.; Gertus, T.; Peckus, M.; Trull, J.; Cojocaru, C.; Sirutkaitis, V.; Staliunas, K. Signatures of light-beam spatial filtering in a three-dimensional photonic crystal. *Phys. Rev. A* **2010**, *82*, 043819. [CrossRef]
61. Wong, S.; Deubel, M.; Pérez-Willard, F.; John, S.; Ozin, G.; Wegener, M.; von Freymann, G. Direct Laser Writing of Three-Dimensional Photonic Crystals with a Complete Photonic Bandgap in Chalcogenide Glasses. *Adv. Mater.* **2006**, *18*, 265–269. [CrossRef]
62. Gailevičius, D.; Purlys, V.; Maigytė, L.; Peckus, M.; Staliūnas, K. Chirped axisymmetric photonic microstructures for spatial filtering. *J. Nanophotonics* **2014**, *8*, 084094. [CrossRef]
63. Purlys, V.; Maigyte, L.; Gailevičius, D.; Peckus, M.; Malinauskas, M.; Gadonas, R.; Staliunas, K. Spatial filtering by axisymmetric photonic microstructures. *Opt. Lett.* **2014**, *39*, 929–932. [CrossRef]
64. Gailevičius, D.; Koliadenko, V.; Purlys, V.; Peckus, M.; Taranenko, V.; Staliunas, K. Photonic crystal microchip laser. *Sci. Rep.* **2016**, *6*, 34173. [CrossRef]
65. Gailevičius, D.; Purlys, V.; Staliunas, K. Photonic crystal spatial filters fabricated by femtosecond pulsed Bessel beam. *Opt. Lett.* **2019**, *44*, 4969–4972. [CrossRef] [PubMed]
66. Davis, K.M.; Miura, K.; Sugimoto, N.; Hirao, K. Writing waveguides in glass with a femtosecond laser. *Opt. Lett.* **1996**, *21*, 1729–1731. [CrossRef] [PubMed]
67. Caulier, O.; Coq, D.; Bychkov, E.; Masselin, P. Direct laser writing of buried waveguide in As_2S_3 glass using a helical sample translation. *Opt. Lett.* **2013**, *38*, 4212. [CrossRef]
68. Long, X.; Bai, J.; Zhao, W.; Stoian, R.; Hui, R.; Cheng, G. Stressed waveguides with tubular depressed-cladding inscribed in phosphate glasses by femtosecond hollow laser beams. *Opt. Lett.* **2012**, *37*, 3138–3140. [CrossRef]
69. Shah, L.; Arai, A.Y.; Eaton, S.M.; Herman, P.R. Waveguide writing in fused silica with a femtosecond fiber laser at 522 nm and 1 MHz repetition rate. *Opt. Express* **2005**, *13*, 1999–2006. [CrossRef] [PubMed]
70. Yang, W.; Corbari, C.; Kazansky, P.G.; Sakaguchi, K.; Carvalho, I.C. Low loss photonic components in high index bismuth borate glass by femtosecond laser direct writing. *Opt. Express* **2008**, *16*, 16215–16226. [CrossRef]
71. Osellame, R.; Taccheo, S.; Marangoni, M.; Ramponi, R.; Laporta, P.; Polli, D.; Silvestri, S.; Cerullo, G. Femtosecond writing of active optical waveguides with astigmatically shaped beams. *J. Opt. Soc. Am. B* **2003**, *20*, 1559–1567. [CrossRef]
72. Liao, Y.; Qi, J.; Wang, P.; Chu, W.; Wang, Z.; Qiao, L.; Cheng, Y. Transverse writing of three-dimensional tubular optical waveguides in glass with a slit-shaped femtosecond laser beam. *Sci. Rep.* **2016**, *6*, 28790. [CrossRef]
73. Nasu, Y.; Kohtoku, M.; Hibino, Y. Low-loss waveguides written with a femtosecond laser for flexible interconnection in a planar light-wave circuit. *Opt. Lett.* **2005**, *30*, 723–725. [CrossRef]
74. Zhang, B.; Xiong, B.; Li, Z.; Li, L.; Lv, J.; Lu, Q.; Wang, L.; Chen, F. Mode tailoring of laser written waveguides in LiNbO3 crystals by multi-scan of femtosecond laser pulses. *Opt. Mater.* **2018**, *86*, 571–575. [CrossRef]
75. Eaton, S.M.; Zhang, H.; Herman, P.R.; Yoshino, F.; Shah, L.; Bovatsek, J.; Arai, A.Y. Heat accumulation effects in femtosecond laser-written waveguides with variable repetition rate. *Opt. Express* **2005**, *13*, 4708–4716. [CrossRef] [PubMed]
76. Zhang, H.; Eaton, S.M.; Li, J.; Herman, P.R. Femtosecond laser direct writing of multiwavelength Bragg grating waveguides in glass. *Opt. Lett.* **2006**, *31*, 3495–3497. [CrossRef] [PubMed]
77. Marshall, G.D.; Ams, M.; Withford, M.J. Direct laser written waveguide-Bragg gratings in bulk fused silica. *Opt. Lett.* **2006**, *31*, 2690–2691. [CrossRef] [PubMed]
78. Osellame, R.; Maselli, V.; Vazquez, R.M.; Ramponi, R.; Cerullo, G. Integration of optical waveguides and microfluidic channels both fabricated by femtosecond laser irradiation. *Appl. Phys. Lett.* **2007**, *90*, 231118. [CrossRef]

79. Vazquez, R.M.; Osellame, R.; Nolli, D.; Dongre, C.; Vlekkert, H.; Ramponi, R.; Pollnau, M.; Cerullo, G. Integration of femtosecond laser written optical waveguides in a lab-on-chip. *Lab Chip* **2009**, *9*, 91–96. [CrossRef] [PubMed]
80. Djogo, G.; Li, J.; Ho, S.; Haque, M.; Ertorer, E.; Liu, J.; Song, X.; Suo, J.; Herman, P.R. Femtosecond laser additive and subtractive micro-processing: Enabling a high-channel-density silica interposer for multicore fibre to silicon-photonic packaging. *Int. J. Extreme Manuf.* **2019**, *1*, 045002. [CrossRef]
81. Bricchi, E.; Klappauf, B.G.; Kazansky, P.G. Form birefringence and negative index change created by femtosecond direct writing in transparent materials. *Opt. Lett.* **2004**, *29*, 119–121. [CrossRef] [PubMed]
82. Sudrie, L.; Franco, M.; Prade, B.; Mysyrowicz, A. Writing of permanent birefringent microlayers in bulk fused silica with femtosecond laser pulses. *Opt. Commun.* **1999**, *171*, 279–284. [CrossRef]
83. Sakakura, M.; Lei, Y.; Wang, L.; Yu, Y.H.; Kazansky, P.G. Ultralow-loss geometric phase and polarization shaping by ultrafast laser writing in silica glass. *Light Sci. Appl.* **2020**, *9*, 15. [CrossRef] [PubMed]
84. Shimotsuma, Y.; Hirao, K.; Qiu, J.; Kazansky, P.G. Nano-modificationinside transparent materials by femtosecond laser single beam. *Mod. Phys. Lett. B* **2005**, *19*, 225–238. [CrossRef]
85. Shimotsuma, Y.; Hirao, K.; Qiu, J.; Miura, K. Nanofabrication in transparent materials with a femtosecond pulse laser. *J. Non-Cryst. Solids* **2006**, *352*, 646–656. [CrossRef]
86. Richter, S.; Miese, C.; Doring, S.; Zimmermann, F.; Withford, M.J.; Tunnermann, A.; Nolte, S. Laser induced nanogratings beyond fused silica—Periodic nanostructures in borosilicate glasses and ULE™. *Opt. Mater. Express* **2013**, *3*, 1161–1166. [CrossRef]
87. Liao, Y.; Ni, J.; Qiao, L.; Huang, M.; Bellouard, Y.; Sugioka, K.; Cheng, Y. High-fidelity visualization of formation of volume nanogratings in porous glass by femtosecond laser irradiation. *Optica* **2015**, *2*, 329–334. [CrossRef]
88. Beresna, M.; Kazansky, P.G. Polarization diffraction grating produced by femtosecond laser nanostructuring in glass. *Opt. Lett.* **2010**, *35*, 1662–1664. [CrossRef]
89. Bricchi, E.; Mills, J.D.; Kazansky, P.G.; Klappauf, B.G.; Baumberg, J.J. Birefringent Fresnel zone plates in silica fabricated by femtosecond laser machining. *Opt. Lett.* **2002**, *27*, 2200–2202. [CrossRef]
90. Papazoglou, D.G.; Loulakis, M.J. Embedded birefringent computer-generated holograms fabricated by femtosecond laser pulses. *Opt. Lett.* **2006**, *31*, 1441–1443. [CrossRef]
91. Cai, W.; Libertun, A.R.; Piestun, R. Polarization selective computer-generated holograms realized in glass by femtosecond laser induced nanogratings. *Opt. Express* **2006**, *14*, 3785–3791. [CrossRef] [PubMed]
92. Beresna, M.; Gecevičius, M.; Kazansky, P.G. Polarization sensitive elements fabricated by femtosecond laser nanostructuring of glass [Invited]. *Opt. Mater. Express* **2011**, *1*, 783–795. [CrossRef]
93. Beresna, M.; Gecevičius, M.; Kazansky, P.G.; Gertus, T. Radially polarized optical vortex converter created by femtosecond laser nanostructuring of glass. *Appl. Phys. Lett.* **2011**, *98*, 201101. [CrossRef]
94. Tsai, Y.J.; Larouche, S.; Tyler, T.; Llopis, A.; Royal, M.; Jokerst, N.M.; Smith, D.R. Arbitrary birefringent metamaterials for holographic optics at λ = 155 μm. *Opt. Express* **2013**, *21*, 26620–26630. [CrossRef] [PubMed]
95. Bricchi, E.; Kazansky, P.G. Extraordinary stability of anisotropic femtosecond direct-written structures embedded in silica glass. *Appl. Phys. Lett.* **2006**, *88*, 111119. [CrossRef]
96. Gallais, L.; Commandré, M. Laser-induced damage thresholds of bulk and coating optical materials at 1030 nm, 500 fs. *Appl. Opt.* **2013**, *53*, A186–A196. [CrossRef] [PubMed]
97. Gu, M.; Li, X. The Road to Multi-Dimensional Bit-by-Bit Optical Data Storage. *Opt. Photonics News* **2010**, *21*, 28–33. [CrossRef]
98. Shimotsuma, Y.; Sakakura, M.; Miura, K.; Qiu, J.; Kazansky, P.G.; Fujita, K.; Hirao, A. Application of Femtosecond-Laser Induced Nanostructures in Optical Memory. *J. Nanosci. Nanotechnol.* **2007**, *7*, 94–104. [CrossRef]
99. Gu, M.; Li, X.; Cao, Y. Optical storage arrays: A perspective for future big data storage. *Light Sci. Appl.* **2014**, *3*, e177. [CrossRef]
100. Beresna, M.; Gecevičius, M.; Kazansky, P.G.; Taylor, T.; Kavokin, A.V. Exciton mediated self-organization in glass driven by ultrashort light pulses. *Appl. Phys. Lett.* **2012**, *101*, 053120. [CrossRef]
101. Zhang, J.; Gecevičius, M.; Beresna, M.; Kazansky, P.G. Seemingly Unlimited Lifetime Data Storage in Nanostructured Glass. *Phys. Rev. Lett.* **2014**, *112*, 033901. [CrossRef]
102. Taylor, R.S.; Hnatovsky, C.; Simova, E.; Rajeev, P.P.; Rayner, D.M.; Corkum, P.B. Femtosecond laser erasing and rewriting of self-organized planar nanocracks in fused silica glass. *Opt. Lett.* **2007**, *32*, 2888–2890. [CrossRef] [PubMed]
103. Chen, F.S.; LaMacchia, J.T.; Fraser, D.B. Holographic storage in lithium niobate. In *Landmark Papers on Photorefractive Nonlinear Optics*; World Scientific: Singapore, 1995; pp. 33–35. [CrossRef]
104. Mizeikis, V.; Purlys, V.; Paipulas, D.; R, B.; Juodkazis, S. Direct Laser Writing: Versatile Tool for Microfabrication of Lithium Niobate. *J. Laser Micro. Nanoeng.* **2012**, *7*, 345–350. [CrossRef]
105. Rusitschka, S.; Curry, E. Big Data in the Energy and Transport Sectors. In *New Horizons for a Data-Driven Economy*; Springer International Publishing: Cham, Switzerland, 2016; pp. 225–244. [CrossRef]
106. Chon, J.W.M.; Gan, X.; Gu, M. Splitting of the focal spot of a high numerical-aperture objective in free space. *Appl. Phys. Lett.* **2002**, *81*, 1576–1578. [CrossRef]
107. Bomzon, Z.; Gu, M.; Shamir, J. Angular momentum and geometrical phases in tight-focused circularly polarized plane waves. *Appl. Phys. Lett.* **2006**, *89*, 241104. [CrossRef]
108. Wyant, J.C.; Creath, K. Basic wavefront aberration theory for optical metrology. *Appl. Opt. Opt. Eng.* **1992**, *11*, 2.

109. Kontenis, G.; Gailevičius, D.; Jonušauskas, L.; Purlys, V. Dynamic aberration correction via spatial light modulator (SLM) for femtosecond direct laser writing: Towards spherical voxels. *Opt. Express* **2020**, *28*, 27850–27864. [CrossRef]
110. Bonora, S.; Jian, Y.; Zhang, P.; Zam, A.; Pugh, E.N.; Zawadzki, R.J.; Sarunic, M.V. Wavefront correction and high-resolution in vivo OCT imaging with an objective integrated multi-actuator adaptive lens. *Opt. Express* **2015**, *23*, 21931–21941. [CrossRef]
111. Li, L.; Wang, D.; Liu, C.; Wang, Q.H. Zoom microscope objective using electrowetting lenses. *Opt. Express* **2016**, *24*, 2931–2940. [CrossRef]
112. Salter, P.S.; Baum, M.; Alexeev, I.; Schmidt, M.; Booth, M.J. Exploring the depth range for three-dimensional laser machining with aberration correction. *Opt. Express* **2014**, *22*, 17644. [CrossRef]
113. de la Cruz, A.R.; Ferrer, A.; Gawelda, W.; Puerto, D.; Sosa, M.G.; Siegel, J.; Solis, J. Independent control of beam astigmatism and ellipticity using a SLM for fs-laser waveguide writing. *Opt. Express* **2009**, *17*, 20853–20859. [CrossRef]
114. Sakakura, M.; Sawano, T.; Shimotsuma, Y.; Miura, K.; Hirao, K. Fabrication of three-dimensional 1 × 4 splitter waveguides inside a glass substrate with spatially phase modulated laser beam. *Opt. Express* **2010**, *18*, 12136–12143. [CrossRef] [PubMed]
115. Simmonds, R.D.; Salter, P.S.; Jesacher, A.; Booth, M.J. Three dimensional laser microfabrication in diamond using a dual adaptive optics system. *Opt. Express* **2011**, *19*, 24122–24128. [CrossRef]
116. Courvoisier, A.; Booth, M.J.; Salter, P.S. Inscription of 3D waveguides in diamond using an ultrafast laser. *Appl. Phys. Lett.* **2016**, *109*, 031109. [CrossRef]
117. Bharadwaj, V.; Courvoisier, A.; Fernandez, T.T.; Ramponi, R.; Galzerano, G.; Nunn, J.; Booth, M.J.; Osellame, R.; Eaton, S.M.; Salter, P.S. Femtosecond laser inscription of Bragg grating waveguides in bulk diamond. *Opt. Lett.* **2017**, *42*, 3451–3453. [CrossRef]
118. Wang, P.; Qi, J.; Liu, Z.; Liao, Y.; Chu, W.; Cheng, Y. Fabrication of polarization-independent waveguides deeply buried in lithium niobate crystal using aberration-corrected femtosecond laser direct writing. *Sci. Rep.* **2017**, *7*, 41211. [CrossRef] [PubMed]
119. Huang, L.; Salter, P.; Karpiński, M.; Smith, B.; Payne, F.; Booth, M. Waveguide fabrication in KDP crystals with femtosecond laser pulses. *Appl. Phys. A* **2014**, *118*, 831–836. [CrossRef]
120. Vizsnyiczai, G.; Kelemen, L.; Ormos, P. Holographic multi-focus 3D two-photon polymerization with real-time calculated holograms. *Opt. Express* **2014**, *22*, 24217–24223. [CrossRef] [PubMed]
121. Horie, Y.; Arbabi, A.; Arbabi, E.; Kamali, S.M.; Faraon, A. High-speed, phase-dominant spatial light modulation with silicon-based active resonant antennas. *ACS Photonics* **2017**, *5*, 1711–1717. [CrossRef]
122. Zhu, G.; Whitehead, D.; Perrie, W.; Allegre, O.J.; Olle, V.; Li, Q.; Tang, Y.; Dawson, K.; Jin, Y.; Edwardson, S.P.; et al. Investigation of the thermal and optical performance of a spatial light modulator with high average power picosecond laser exposure for materials processing applications. *J. Phys. D Appl. Phys.* **2018**, *51*, 095603. [CrossRef]
123. Park, J.; Jeong, B.G.; Kim, S.I.; Lee, D.; Kim, J.; Shin, C.; Lee, C.B.; Otsuka, T.; Kyoung, J.; Kim, S.; et al. All-solid-state spatial light modulator with independent phase and amplitude control for three-dimensional LiDAR applications. *Nature Nanotechnol.* **2020**, *16*, 69–76. [CrossRef] [PubMed]
124. Marcinkevičius, A.; Juodkazis, S.; Watanabe, M.; Miwa, M.; Matsuo, S.; Misawa, H.; Nishii, J. Femtosecond laser-assisted three-dimensional microfabrication in silica. *Opt. Lett.* **2001**, *26*, 277–279. [CrossRef] [PubMed]
125. Kiyama, S.; Matsuo, S.; Hashimoto, S.; Morihira, Y. Examination of Etching Agent and Etching Mechanism on Femotosecond Laser Microfabrication of Channels Inside Vitreous Silica Substrates. *J. Phys. Chem. C* **2009**, *113*, 11560–11566. [CrossRef]
126. Ross, C.A.; Maclachan, D.G.; Choudrury, D.; Thomson, R.R. Optimisation of ultrafast laser assisted etching in fused silica. *Opt. Express* **2018**, *26*, 24343–24356. [CrossRef] [PubMed]
127. Chen, F.; Deng, Z.; Yang, Q.; Bian, H.; Du, G.; Si, J.; Hou, X. Rapid fabrication of a large-area close-packed quasi-periodic microlens array on BK7 glass. *Opt. Lett.* **2014**, *39*, 606–609. [CrossRef]
128. Matsuo, S.; Sumi, H.; Kiyama, S.; Tomita, T.; Hashimoto, S. Femtosecond laser-assisted etching of Pyrex glass with aqueous solution of KOH. *Appl. Surf. Sci.* **2009**, *225*, 9758–9760. [CrossRef]
129. Juodkazis, S.; Yamasaki, K.; Mizeikis, V.; Matsuo, S.; Misawa, H. Formation of embedded patterns in glasses using femtosecond irradiation. *Appl. Phys. A* **2004**, *79*, 1549–1553. [CrossRef]
130. Hu, Y.; Rao, S.; Wu, S.; Wei, P.; Qiu, W.; Wu, D.; Xu, B.; Ni, J.; Yang, L.; Li, J.; et al. All-Glass 3D Optofluidic Microchip with Built-in Tunable Microlens Fabricated by Femtosecond Laser-Assisted Etching. *Adv. Opt. Mater.* **2018**, *6*, 1701299. [CrossRef]
131. Butkus, S.; Rickus, M.; Sirutkaitis, R.; Paipulas, D.; Sirutkaitis, V. Fabrication of High Aspect Ratio Channels in Fused Silica Using Femtosecond Pulses and Chemical Etching at Different Conditions. *J. Laser Micro. Nanoen.* **2019**, *14*, 19–24.
132. LoTurco, S.; Osellame, R.; Ramponi, R.; Vishnubhatla, K.C. Hybrid chemical etching of femtosecond laser irradiated structures for engineered microfluidic devices. *J. Micromech. Microeng.* **2013**, *23*, 1–8. [CrossRef]
133. Bu, M.; Melvin, T.; Ensell, G.J.; Wilkinson, J.S.; Evans, A.G. A new masking technology for deep glass etching and its microfluidic application. *Sens. Actuators A* **2004**, *115*, 476–482. [CrossRef]
134. Venturini, F.; Navarrini, W.; Resnati, G.; Metrangolo, P.; Vazquez, R.M.; Osellame, R.; Cerullo, G. Selective Iterative Etching of Fused Silica with Gaseous Hydrofluoric Acid. *J. Phys. Chem. C* **2010**, *114*, 18712–18716. [CrossRef]
135. LoTurco, S. Hybrid chemical etching of femtosecond irradiated 3D structures in fused silica glass. *MATEC Web Conf.* **2013**, *8*, 05009. [CrossRef]
136. Choudhury, D.; Ródenas, A.; Paterson, L.; Jaque, D.; Kar, A.K. 3D Microfabrication in YAG Crystals by Direct Laser Writing and Chemical Etching. In Proceedings of the Lasers and Electro-Optics Pacific Rim Conference (Optical Society of America, 2013), Kyoto, Japan, 30 June–4 July 2013; pp. 14.

137. Paiè, P.; Bragheri, F.; Carlo, D.D.; Osellame, R. Particle focusing by 3D inertial microfluidics. *Microsyst. Nanoeng.* **2017**, *3*. [CrossRef] [PubMed]
138. Randles, A.B.; Esashi, M.; Tanaka, S. Etch rate dependence on crystal orientation of lithium niobate. *IEEE Trans. Ultrason. Ferroelectr. Freq. Control.* **2010**, *57*, 2372–2380. [CrossRef] [PubMed]
139. Matsuo, S.; Tabuchi, Y.; Okada, T.; Juodkazis, S.; Misawa, H. Femtosecond laser assisted etching of quartz: Microstructuring from inside. *Appl. Phys. A* **2006**, *84*, 99–102. [CrossRef]
140. Wortmann, D.; Gottmann, J.; Brandt, N.; Horn-Solle, H. Micro- and nanostructures inside sapphire by fs-laser irradiation and selective etching. *Opt. Express* **2008**, *16*, 1517–1522. [CrossRef]
141. Gottmann, J.; Wortmann, D.; Horstmann-Jungemann, M. Fabrication of sub-wavelength surface ripples and in-volume nanostructures by fs-laser induced selective etching. *Appl. Surf. Sci.* **2009**, *255*, 5641–5646. [CrossRef]
142. Horstmann-Jungemann, M.; Gottmann, J.; Wortmann, D. Nano- and Microstructuring of SiO_2 and Sapphire with Fs-laser Induced Selective Etching. *J. Laser. Micro. Nanoeng.* **2009**, *4*, 135–140. [CrossRef]
143. Juodkazis, S.; Nishimura, K.; Tanaka, S.; Misawa, H.; Gamaly, E.G.; Luther-Davies, B.; Hallo, L.; Nicolai, P.; Tikhonchuk, V.T. Laser-Induced Microexplosion Confined in the Bulk of a Sapphire Crystal: Evidence of Multimegabar Pressures. *Phys. Rev. Lett.* **2006**, *96*, 166101. [CrossRef]
144. Hörstmann-Jungemann, M.; Gottmann, J.; Keggenhoff, M. 3D-Microstructuring of Sapphire using fs-Laser Irradiation and Selective Etching. *J. Laser. Micro. Nanoeng.* **2010**, *5*, 145–149. [CrossRef]
145. Capuano, L.; Pohl, R.; Tiggelaar, R.M.; Berenschot, J.W.; Gardeniers, J.E.; RÖMer, G.R.B.E. Morphology of single picosecond pulse subsurface laser-induced modifications of sapphire and subsequent selective etching. *Opt. Express* **2018**, *26*, 29283–29295. [CrossRef] [PubMed]
146. Capuanoa, L.; Tiggelaar, R.; Berenschot, J.; Gardeniers, J.; Römer, G. Fabrication of millimeter-long structures in sapphire using femtosecond infrared laser pulses and selective etching. *Opt. Laser Eng.* **2020**, *133*, 106114. [CrossRef]
147. Juodkazis, S.; Nishi, Y.; Misawa, H. Femtosecond laser-assisted formation of channels in sapphire using KOH solution. *Phys. Status Solidi RRL* **2008**, *2*, 275–277. [CrossRef]
148. Li, Q.; Chen, Q.; Niu, L.; Yu, Y.; Wang, L.; Sun, Y.; Sun, H. Sapphire-Based Dammann Gratings for UV Beam Splitting. *IEEE Photon. J.* **2016**, *8*, 2500208. [CrossRef]
149. Li, Q.; Yu, Y.; Wang, L.; Cao, X.; Liu, X.; Sun, Y.; Chena, Q.; Duan, J.; Sun, H. Sapphire-Based Fresnel Zone Plate Fabricated by Femtosecond Laser Direct Writing and Wet Etching. *IEEE Photon. Technol. Lett.* **2016**, *28*, 1290–1293. [CrossRef]
150. Matsuo, S.; Tokumi, K.; Tomita, T.; Hashimoto, S. Three-Dimensional Residue-Free Volume Removal inside Sapphire by High-Temperature Etching after Irradiation of Femtosecond Laser Pulses. *Laser Chem.* **2008**, *2008*, 892721. [CrossRef]
151. Hermans, M.; Gottmann, J.; Riedel, F. Selective, Laser-Induced Etching of Fused Silica at High Scan-Speeds Using KOH. *J. Laser. Micro. Nanoeng.* **2014**, *9*, 126–131. [CrossRef]
152. Li, X.; Xu, J.; Lin, Z.; Qi, J.; Wang, P.; Chu, W.; Fang, Z.; Wang, Z.; Chai, Z.; Cheng, Y. Polarization-insensitive space-selective etching in fused silica induced by picosecond laser irradiation. *Appl. Surf. Sci.* **2019**, *485*, 188–193. [CrossRef]
153. Sugioka, K.; Masuda, M.; Hongo, T.; Cheng, Y.; Shihoyama, K.; Midorikawa, K. Three-dimensional microfluidic structure embedded in photostructurable glass by femtosecond laser for lab-on-chipapplications. *Appl. Phys. A* **2004**, *79*, 815–817. [CrossRef]
154. Gottmann, J.; Hermans, M.; Repiev, N.; Ortmann, J. Selective Laser-Induced Etching of 3D Precision Quartz Glass Components for Microfluidic Applications-Up-Scaling of Complexity and Speed. *Micromachines* **2017**, *8*, 110. [CrossRef]
155. Kim, S.; Kim, J.; Joung, Y.H.; Ahn, S.; Choi, J.; Koo, C. Optimization of selective laser-induced etching (SLE) for fabrication of 3D glass microfluidic device with multi-layer micro channels. *Micro Nano Sys. Lett.* **2019**, *7*, 15. [CrossRef]
156. Sugioka, K.; Cheng, Y.; Midorikawa, K. Three-dimensional micromachining of glass using femtosecond laser for lab-on-a-chip device manufacture. *Appl. Phys. A* **2005**, *81*, 1–10. [CrossRef]
157. Bellouard, Y.; Lehnert, T.; Bidaux, J.E.; Sidler, T.; Clavel, R.; Gotthardt, R. Local annealing of complex mechanical devices: A new approach for developing monolithic micro-devices. *Mater. Sci. Eng.* **1999**, *273*, 795–798. [CrossRef]
158. Bellouard, Y. Shape memory alloys for microsystems: A review from a material research perspective. *Mater. Sci. Eng. A* **2008**, *481*, 582–589. [CrossRef]
159. Gottmann, J.; Hermans, M.; ortmann, J. Microcutting and Hollow 3D Microstructures in Glasses by In-volume Selective Laser-induced Etching (ISLE). *J. Laser. Micro. Nanoeng.* **2013**, *8*, 15–18. [CrossRef]
160. Tičkūnas, T.; Perrenoud, M.; Butkus, S.; Gadonas, R.; Rekštytė, S.; Malinauskas, M.; Paipulas, D.; Bellouard, Y.; Sirutkaitis, V. Combination of additive and subtractive laser 3D microprocessing in hybrid glass/polymer microsystems for chemical sensing applications. *Opt. Express* **2017**, *25*, 26280–26288. [CrossRef]
161. Wu, D.; Xu, J.; Niu, L.G.; Wu, S.Z.; Midorikawa, K.; Sugioka, K. In-channel integration of designable microoptical devices using flat scaffold-supported femtosecond-laser microfabrication for coupling-free optofluidic cell counting. *Light Sci. Appl.* **2015**, *4*, e228. [CrossRef]
162. Wu, D.; Niu, L.G.; Wu, S.Z.; Xu, J.; Midorikawa, K.; Sugioka, K. Ship-in-a-bottle femtosecond laser integration of optofluidic microlens arrays with center-pass units enabling coupling-free parallel cell counting with a 100% success rate. *Lab Chip* **2015**, *15*, 1515–1523. [CrossRef]

163. Bückmann, T.; Stenger, N.; Kadic, M.; Kaschke, J.; Frölich, A.; Kennerknecht, T.; Eberl, C.; Thiel, M.; Wegener, M. Tailored 3D mechanical metamaterials made by dip-in direct-laser-writing optical lithography. *Adv. Mater.* **2012**, *24*, 2710–2714. [CrossRef]
164. Jonušauskas, L.; Baravykas, T.; Andrijec, D.; Gadišauskas, T.; Purlys, V. Stitchless support-free 3D printing of free-form micromechanical structures with feature size on-demand. *Sci. Rep.* **2019**, *9*, 17533. [CrossRef]
165. Yang, D.; Liu, L.; Gong, Q.; Li, Y. Rapid Two-Photon Polymerization of an Arbitrary 3D Microstructure with 3D Focal Field Engineering. *Macromol. Rapid Commun.* **2019**, *40*, 1900041. [CrossRef]
166. Dubey, A.K.; Yadava, V. Laser beam machining—A review. *Int. J. Mach. Tools Manuf.* **2008**, *48*, 609–628. [CrossRef]
167. Garškaitė, E.; Alinauskas, L.; Drienovsky, M.; Krajcovic, J.; Cicka, R.; Palcut, M.; Jonušauskas, L.; Malinauskas, M.; Stankeviciute, Z.; Kareiva, A. Fabrication of a composite of nanocrystalline carbonated hydroxyapatite (cHAP) with polylactic acid (PLA) and its surface topographical structuring with direct laser writing (DLW). *RSC Adv.* **2016**, *6*, 72733–72743. [CrossRef]
168. Ye, G.; Wang, W.; Fan, D.; He, P. Effects of femtosecond laser micromachining on the surface and substrate properties of poly-lactic acid (PLA). *Appl. Surf. Sci.* **2020**, *538*, 148117. [CrossRef]
169. Binder, P.S. Flap dimensions created with the IntraLase FS laser. *J. Cataract. Refract. Surg.* **2004**, *30*, 26–32. [CrossRef]
170. Kruger, J.; Kautek, W.; Lenzner, M.; Sartania, S.; Spielmann, C.; Krausz, F. Laser micromachining of barium aluminium borosilicate glass with pulse durations between 20 fs and 3 ps. *Appl. Surf. Sci.* **1998**, *127–129*, 892–898. [CrossRef]
171. Ben-Yakar, A.; Byer, R.L. Femtosecond laser ablation properties of borosilicate glass. *J. Appl. Phys.* **2004**, *96*, 5316–5323. [CrossRef]
172. Xu, S.; Qiu, J.; Jia, T.; Li, C.; Sun, H.; Xu, Z. Femtosecond laser ablation of crystals SiO_2 and YAG. *Opt. Commun.* **2007**, *274*, 163–166. [CrossRef]
173. Furusawa, K.; Takahashi, K.; Kumagai, H.; Midorikawa, K.; Obara, M. Ablation characteristics of Au, Ag, and Cu metals using a femtosecond Ti:sapphire laser. *Appl. Phys. A* **1999**, *69*, S359–S366. [CrossRef]
174. Žemaitis, A.; Gaidys, M.; Gečys, P.; Račiukaitis, G.; Gedvilas, M. Rapid high-quality 3D micro-machining by optimised efficient ultrashort laser ablation. *Opt. Lasers Eng.* **2018**, *114*, 83–89. [CrossRef]
175. Gafner, M.; Kramer, T.; Remund, S.M.; Holtz, R.; Neuenschwander, B. Ultrafast pulsed laser high precision micromachining of rotational symmetric parts. *J. Laser Appl.* **2021**, *33*, 012053. [CrossRef]
176. Giridhar, M.S.; Seong, K.; Schülzgen, A.; Khulbe, P.; Peyghambarian, N.; Mansuripur, M. Femtosecond pulsed laser micromachining of glass substrates with application to microfluidic devices. *Appl. Opt.* **2004**, *43*, 4584–4589. [CrossRef]
177. Nikumb, S.; Chen, Q.; Li, C.; Reshef, H.; Zheng, H.; Qiu, H.; Low, D. Precision glass machining, drilling and profile cutting by short pulse lasers. *Thin Solid Films* **2005**, *477*, 216–221. [CrossRef]
178. Queste, S.; Salut, R.; Clatot, S.; Rauch, J.Y.; Malek, C.G.K. Manufacture of microfluidic glass chips by deep plasma etching, femtosecond laser ablation, and anodic bonding. *Microsyst. Technol.* **2010**, *16*, 1485–1493. [CrossRef]
179. Ke, K.; Hasselbrink, E.F.; Hunt, A.J. Rapidly Prototyped Three-Dimensional Nanofluidic Channel Networks in Glass Substrates. *Anal. Chem.* **2005**, *77*, 5083–5088. [CrossRef] [PubMed]
180. Li, Y.; Qu, S.; Guo, Z. Fabrication of microfluidic devices in silica glass by water-assisted ablation with femtosecond laser pulses. *J. Micromech. Microeng.* **2011**, *21*, 075008. [CrossRef]
181. Jonušauskas, L.; Rekštytė, S.; Buividas, R.; Butkus, S.; Gadonas, R.; Juodkazis, S.; Malinauskas, M. Hybrid Subtractive-Additive-Welding Microfabrication for Lab-on-Chip (LOC) Applications via Single Amplified Femtosecond Laser Source. *Opt. Eng.* **2017**, *56*, 094108. [CrossRef]
182. Bellouard, Y.; Hongler, M.O. Femtosecond-laser generation of self-organized bubble patterns in fused silica. *Opt. Express* **2011**, *19*, 6807–6821. [CrossRef]
183. Li, Y.; Itoh, K.; Watanabe, W.; Yamada, K.; Kuroda, D.; Nishii, J.; Jiang, Y. Three-dimensional hole drilling of silica glass from the rear surface with femtosecond laser pulses. *Opt. Lett.* **2001**, *26*, 1912. [CrossRef]
184. Bhuyan, M.K.; Courvoisier, F.; Lacourt, P.A.; Jacquot, M.; Furfaro, L.; Withford, M.J.; Dudley, J.M. High aspect ratio taper-free microchannel fabrication using femtosecond Bessel beams. *Opt. Express* **2010**, *18*, 566–574. [CrossRef]
185. Liao, Y.; Ju, Y.; Zhang, L.; He, F.; Zhang, Q.; Shen, Y.; Chen, D.; Cheng, Y.; Xu, Z.; Sugioka, K.; et al Three-dimensional microfluidic channel with arbitrary length and configuration fabricated inside glass by femtosecond laser direct writing. *Opt. Lett.* **2010**, *35*, 3225. [CrossRef]
186. Sugioka, K.; Hanada, Y.; Midorikawa, K. Three-dimensional femtosecond laser micromachining of photosensitive glass for biomicrochips. *Laser Photonics Rev.* **2010**, *4*, 386–400. [CrossRef]
187. Ju, Y.; Liao, Y.; Zhang, L.; Sheng, Y.; Zhang, Q.; Chen, D.; Cheng, Y.; Xu, Z.; Sugioka, K.; Midorikawa, K. Fabrication of large-volume microfluidic chamber embedded in glass using three-dimensional femtosecond laser micromachining. *Microfluid. Nanofluid.* **2011**, *11*, 111–117. [CrossRef]
188. Liao, Y.; Song, J.; L., E.; Luo, Y.; Shen, Y.; Chen, D.; Cheng, Y.; Xu, Z.; Sugioka, K.; Midorikawa, K. Rapid prototyping of three-dimensional microfluidic mixers in glass by femtosecond laser direct writing. *Lab Chip* **2012**, *12*, 746–749. [CrossRef] [PubMed]
189. Gissibl, T.; Thiele, S.; Herkommer, A.; Giessen, H. Two-photon direct laser writing of ultracompact multi-lens objectives. *Nature Photon.* **2016**, *10*, 554–560. [CrossRef]
190. Dietrich, P.I.; Blaicher, M.; Reuter, I.; Billah, M.; Hoose, T.; Hofmann, A.; Caer, C.; Dangel, R.; Offrein, B.; Troppenz, U.; et al. In situ 3D nanoprinting of free-form coupling elements for hybrid photonic integration. *Nature Photon.* **2018**, *12*, 241–247. [CrossRef]

191. Hahn, V.; Kalt, S.; Sridharan, G.M.; Wegener, M.; Bhattacharya, S. Polarizing beam splitter integrated onto an optical fiber facet. *Opt. Express* **2018**, *26*, 33148–33157. [CrossRef] [PubMed]
192. Choi, H.K.; Ryu, J.; Kim, C.; Noh, Y.C.; Sohn, I.B.; Kim, J.T. Formation of Micro-lens Array using Femtosecond and CO_2 lasers. *J. Laser Micro. Nanoeng.* **2016**, *11*, 341–345.
193. Schwarz, S.; Hellmann, R. Fabrication of Cylindrical Lenses by CombiningUltrashort Pulsed Laser and CO_2 Laser. *J. Laser Micro. Nanoeng.* **2017**, *12*, 76–79. [CrossRef]
194. Schwarz, S.; Gotzendorfer, B.; Rung, S.; Esen, C.; Hellmann, R. Compact Beam Homogenizer Module with Laser-Fabricated Lens-Arrays. *Appl. Sci.* **2021**, *11*, 1018. [CrossRef]
195. Dudutis, J.; Pipiras, J.; Schwarz, S.; Rung, S.; Hellmann, R.; Račiukaitis, G.; Gečys, P. Laser-fabricated axicons challenging the conventional optics in glass processing applications. *Opt. Express* **2020**, *28*, 5715–5730. [CrossRef]
196. Lin, C.H.; Jiang, L.; Chai, Y.H.; Xiao, H.; Chen, S.J.; Tsai, H.L. Fabrication of microlens arrays in photosensitive glass by femtosecond laser direct writing. *Appl. Phys. A* **2009**, *97*, 751–757. [CrossRef]
197. Pan, A.; Gao, B.; Chen, T.; Si, J.; Li, C.; Chen, F.; Hou, X. Fabrication of concave spherical microlenses on silicon by femtosecond laser irradiation and mixed acid etching. *Opt. Express* **2014**, *22*, 15245–15250. [CrossRef] [PubMed]
198. Zhang, F.; Wang, C.; Yin, K.; Dong, X.R.; Song, Y.X.; Tian, Y.X.; Duan, J.A. Quasi-periodic concave microlens array for liquid refractive index sensing fabricated by femtosecond laser assisted with chemical etching. *Sci. Rep.* **2018**, *8*, 2419. [CrossRef] [PubMed]
199. Weingarten, C.; Schmickler, A.; Willenborg, E.; Wissenbach, K.; Poprawe, R. Laser polishing and laser shape correction of optical glass. *J. Laser Appl.* **2017**, *29*, 011702. [CrossRef]
200. Martinez, S.; Lamikiz, A.; Ukar, E.; Calleja, A.; Arrizubieta, J.A.; de Lacalle, L.N.L. Analysis of the regimes in the scanner-based laser hardening process. *Opt. Lasers Eng.* **2017**, *90*, 72–80. [CrossRef]
201. Weingarten, C.; Steenhusen, S.; Hermans, M.; Willenborg, E.; Schleifenbaum, J.H. Laser polishing and 2PP structuring of inside microfluidic channels in fused silica. *Microfluid. Nanofluid* **2017**, *21*. [CrossRef]
202. Serhatlioglu, M.; Ortaç, B.; Elbuken, C.; Biyikli, N.; Solmaz, M.E. CO_2 laser polishing of microfluidic channels fabricated by femtosecond laser assisted carving. *J. Micromech. Microeng.* **2016**, *26*, 115011. [CrossRef]
203. Xu, G.; Dai, Y.; Cui, J.; Xiao, X.; Mei, H.; Li, H. Simulation and experiment of femtosecond laser polishing quartz material. *Integr. Ferroelectr.* **2017**, *181*, 60–69. [CrossRef]
204. Taylor, L.L.; Xu, J.; Pomerantz, M.; Smith, T.R.; Lambropoulos, J.C.; Qiao, J. Femtosecond laser polishing of germanium. *Opt. Mater. Express* **2019**, *9*, 4165–4177. [CrossRef]
205. Fan, Z.; Sun, X.; Zhuo, X.; Mei, X.; Cui, J.; Duan, W.; Wang, W.; Zhang, X.; Yang, L. Femtosecond laser polishing yttria-stabilized zirconia coatings for improving molten salts corrosion resistance. *Corros. Sci.* **2021**, *184*, 109367. [CrossRef]
206. Mills, B.; Heath, D.J.; Feinaeugle, M.; Grant-Jacob, J.A.; Eason, R.W. Laser ablation via programmable image projection for submicron dimension machining in diamond. *J. Laser Appl.* **2014**, *26*, 041501. [CrossRef]
207. Courvoisier, F.; Zhang, J.; Bhuyan, M.K.; Jacquot, M.; Dudley, J.M. Applications of femtosecond Bessel beams to laser ablation. *Appl.Phys. A* **2012**, *112*, 29–34. [CrossRef]
208. Kim, H.Y.; Yoon, J.W.; Choi, W.S.; Kim, K.R.; Cho, S.H. Ablation depth control with 40 nm resolution on ITO thin films using a square, flat top beam shaped femtosecond NIR laser. *Opt. Lasers Eng.* **2016**, *84*, 44–50. [CrossRef]
209. Hernandez-Rueda, J.; Götte, N.; Siegel, J.; Soccio, M.; Zielinski, B.; Sarpe, C.; Wollenhaupt, M.; Ezquerra, T.A.; Baumert, T.; Solis, J. Nanofabrication of Tailored Surface Structures in Dielectrics Using Temporally Shaped Femtosecond-Laser Pulses. *ACS Appl. Mater. Interfaces* **2015**, *7*, 6613–6619. [CrossRef] [PubMed]
210. Singh, S.; Argument, M.; Tsui, Y.Y.; Fedosejevs, R. Effect of ambient air pressure on debris redeposition during laser ablation of glass. *J. Appl. Phys.* **2005**, *98*, 113520. [CrossRef]
211. Gerhard, C.; Roux, S.; Brückner, S.; Wieneke, S.; Viöl, W. Low-temperature atmospheric pressure argon plasma treatment and hybrid laser-plasma ablation of barite crown and heavy flint glass. *Appl. Opt.* **2012**, *51*, 3847–3852. [CrossRef]
212. Butkus, S.; Gaižauskas, E.; Paipulas, D.; Viburys, Ž.; Kaškelyė, D.; Barkauskas, M.; Alesenkov, A.; Sirutkaitis, V. Rapid microfabrication of transparent materials using filamented femtosecond laser pulses. *Appl. Phys. A* **2013**, *114*, 81–90. [CrossRef]
213. Pan, Y.J.; Yang, R.J. A glass microfluidic chip adhesive bonding method at room temperature. *J. Micrromech. Microeng.* **2006**, *16*, 2666–2672. [CrossRef]
214. Haisma, J.; Hattu, N.; Pulles, J.T.; Steding, E.; Vervest, J.C. Direct bonding and beyond. *Appl. Opt.* **2007**, *46*, 6793–6803. [CrossRef]
215. Tamaki, T.; Watanabe, W.; Nishii, J.; Itoh, K. Welding of transparent materials using femtosecond laser pulses. *Jpn. J. Appl. Phys. Part 2* **2005**, *44*, L687–L689. [CrossRef]
216. Tamaki, T.; Watanabe, W.; Itoh, K. Laser micro-welding of transparent materials by a localized heat accumulation effect using a femtosecond fiber laser at 1558 nm. *Opt. Express* **2006**, *14*, 10460–10468. [CrossRef] [PubMed]
217. Richter, S.; Nolte, S.; Tunnermann, A. Ultrashort pulse laser welding—A new approach for high-stability bonding of different glass. *Phys. Procedia* **2012**, *39*, 556–562. [CrossRef]
218. Richter, S.; Zimmermann, F.; Shutter, D.; Budnicki, A.; Tunnermann, A.; Nolte, S. Ultrashort pulse induced laser wlding of glasses without optical contacting. In Proceedings of the SPIE LASE, San Francisco, CA, USA, 28 January–2 February 2017; Volume 10094, p. 1009411.

219. Helie, D.; Begina, M.; Lacroix, F.; Vallee, R. Reinforced direct bonding of optical materials by femtosecond laser welding. *Appl. Opt.* **2012**, *51*, 2098–2106. [CrossRef] [PubMed]
220. Chen, J.; Carter, R.M.; Thomson, R.R.; Hand, D.P. Avoiding the requirement for pre-existing optical contact during picosecond laser glass-to-glass welding. *Opt. Express* **2015**, *23*, 18645–18657. [CrossRef] [PubMed]
221. Watanabe, W.; Onda, S.; Tamaki, T.; Itoh, K.; Nishii, J. Space-selective laser joining of dissimilar transparent materials using femtosecond laser pulses. *Appl. Phys. Lett.* **2006**, *89*, 021106. [CrossRef]
222. Zhang, G.; Stoian, R.; Zhao, W.; Cheng, G. Femtosecond laser Bessel beam welding of transparent to non-transparent materials with large focal-position tolerant zone. *Opt. Express* **2018**, *26*, 917–926. [CrossRef]
223. Greco, V.; Marchesini, F.; Molesini, G. Optical contact and Van der Waals interactions: The role of the surface topography in determining the bonding strength of thick glass plates. *J. Opt. A Pure Appl. Opt.* **2001**, *3*, 85–88. [CrossRef]
224. Cvecek, K.; Odato, R.; Dehmel, S.; Miyamoto, I.; Schmidt, M. Gap bridging in joining of glass using ultrashort laser pulses. *Opt. Exp.* **2015**, *23*, 5681–5693. [CrossRef] [PubMed]
225. Miyamoto, I.; Cvecek, K.; Schmidt, M. Crack-free conditions in welding of glass by ultrashort laser pulse. *Opt. Express* **2013**, *21*, 14291–14302. [CrossRef] [PubMed]
226. Miyamoto, I.; Cvecek, K.; Okamoto, Y.; Schmidt, M. Novel fusion welding technology of glass sing ultrashort pulse lasers. *Phys. Procedia* **2010**, *5*, 482–493. [CrossRef]
227. Shimizu, M.; Sakakura, M.; Ohnishi, M.; Shimotsuma, Y.; Nakaya, T.; Miura, K.; Hirao, K. Mechanism of heat-modification inside a glass after irradiation with high-repetition rate femtosecond laser pulses. *J. Appl. Phys.* **2010**, *108*, 073533. [CrossRef]
228. Huang, H.; Yang, L.M.; Liu, J. Ultrashort pulsed fiber laser welding and sealing of transparent materials. *Appl. Opt.* **2012**, *51*, 2979–2986. [CrossRef]
229. Zimmermann, F.; Richter, S.; Döring, S.; Tünnermann, A.; Nolte, S. Ultrastable bonding of glass with femtosecond laser bursts. *Appl. Opt.* **2013**, *52*, 1149–1154. [CrossRef] [PubMed]
230. Carter, R.M.; Troughton, M.; Chen, J.; Elder, I.; Thomson, R.R.; Esser, M.J.D.; Lamb, R.A.; Hand, D.P. Towards industrial ultrafast laser microwelding: SiO_2 and BK7 to aluminum alloy. *Appl. Opt.* **2017**, *56*, 4873–4881. [CrossRef] [PubMed]

Article

Effects of Thermal Annealing on Femtosecond Laser Micromachined Glass Surfaces

Federico Sala [1,2], Petra Paié [2,*], Rebeca Martínez Vázquez [2], Roberto Osellame [1,2] and Francesca Bragheri [2]

[1] Department of Physics, Politecnico di Milano, Piazza Leonardo da Vinci 32, 20133 Milano, Italy; federico.sala@polimi.it (F.S.); roberto.osellame@cnr.it (R.O.)
[2] Istituto di Fotonica e Nanotecnologie, CNR, Piazza Leonardo da Vinci 32, 20133 Milano, Italy; rebeca.martinez@polimi.it (R.M.V.); francesca.bragheri@ifn.cnr.it (F.B.)
* Correspondence: petra.paie@cnr.it

Abstract: Femtosecond laser micromachining (FLM) of fused silica allows for the realization of three-dimensional embedded optical elements and microchannels with micrometric feature size. The performances of these components are strongly affected by the machined surface quality and residual roughness. The polishing of 3D buried structures in glass was demonstrated using different thermal annealing processes, but precise control of the residual roughness obtained with this technique is still missing. In this work, we investigate how the FLM irradiation parameters affect surface roughness and we characterize the improvement of surface quality after thermal annealing. As a result, we achieved a strong roughness reduction, from an average value of 49 nm down to 19 nm. As a proof of concept, we studied the imaging performances of embedded mirrors before and after thermal polishing, showing the capacity to preserve a minimum feature size of the reflected image lower than 5 µm. These results allow for us to push forward the capabilities of this enabling fabrication technology, and they can be used as a starting point to improve the performances of more complex optical elements, such as hollow waveguides or micro-lenses.

Keywords: femtosecond laser micromachining; fused silica; roughness analysis; thermal annealing; integrated optics

Citation: Sala, F.; Paié, P.; Martínez Vázquez, R.; Osellame, R.; Bragheri, F. Effects of Thermal Annealing on Femtosecond Laser Micromachined Glass Surfaces. *Micromachines* **2021**, *12*, 180. https://doi.org/10.3390/mi12020180

Academic Editor: Martin Byung-Guk Jun

Received: 12 December 2020
Accepted: 7 February 2021
Published: 11 February 2021

1. Introduction

Femtosecond laser micromachining (FLM) is a versatile technique that allows for the microstructuring of different types of materials. A laser source, with a pulse duration ranging from few tens to many hundreds of femtoseconds, is focused inside of a transparent material. Inside the focal volume, thanks to the high intensity, nonlinear phenomena occur, permanently modifying the substrate. It is possible to move the focal point inside the volume in the three dimensions, laser-writing the desired geometry inside the material. The most versatile class of materials that can be employed for FLM laser writing is glass. Depending on the type of glass and on the irradiation parameters, it is possible to induce different types of modifications. A first type of modification, shared by many types of glasses, is a local permanent change of the refractive index that can be used to realize waveguides and integrated photonic circuits [1]. A second type of modification, characterized by a strong birefringence, can be induced in certain glasses [2–6] by changing the irradiation parameters. This type of modification takes the name of *nanogratings*, and it is characterized by periodic nanostructuring of the laser track in the form of lamellae, usually oriented perpendicularly to the femtosecond laser linear polarization. These nanogratings express a highly enhanced etching rate in specific acids, like hydrofluoric acid (HF). This opens the possibility to selectively microstructure the glass, realizing embedded channels and cavities. The combination of laser irradiation and acid attack takes the name FLICE (femtosecond laser irradiation followed by chemical etching) [7]. An example of glass that presents both types of modifications, depending on the writing laser fluence [8], is fused silica. Its high transparency, mechanical robustness, and chemical inertia combined with the possibility to realize both microstructures and photonic elements make fused silica an optimal platform for the realization of integrated lab-on-chip (LOC) devices in the field of integrated photonics [1], optomechanics [9], and optofluidics [10,11].

The two main building blocks of LOCs are microchannels and micro-optics. Embedded channels are used in microfluidic applications to confine the fluids in a laminar regime and to perform many types of operations such as fluid mixing or filtration, and particle focusing, sorting, delivering, and handling [12]. With FLICE, they can be realized directly inside the volume, or they can have an open geometry, sealed afterwards. A first example of integrated optical components fabricated by FLICE is given by embedded mirrors that make use of total internal reflection (TIR) [13–15] or filled with metallic media [16]. A second example is the realization of focusing elements such as micro-lenses [17], hollow micro-lenses [15], filled micro-lenses [18], or diffraction-based elements [19]. These two components are highly sensitive to a specific manufacturing parameter that can strongly affect their performances: the residual surface roughness. Rough microchannel surfaces can affect the quality of the images taken when looking with a microscope inside the channel. On the other hand, the roughness strongly affects the optics performances, both in the case of light scattering, total internal reflection [20], or light confinement. As a consequence, control over the residual surface roughness is a main issue in FLICE fabrications.

From the literature, the FLICE residual roughness (root mean squared (RMS)) in the case of fused silica is around 60 nm and it strongly depends on laser irradiation polarization [21,22] and scan direction, i.e., on whether the surface is parallel or orthogonal to the writing laser propagation axis [23]. In order to improve the surface quality, several treatments have been proposed. A first example is local thermal annealing of exposed surfaces, performed with oxyhydrogen flame. This procedure causes partial melting of the most superficial layers and subsequent smoothing thanks to surface tension [17,24]. It can be combined with glass stretching [25]. This approach was used to smooth a buried microchannel, but it has a major drawback. The structures are significantly elongated along the drawing direction; thus, this technique is not suitable for complex geometries in which the shape of the microstructures should be preserved, such as in the case of integrated lenses. Another technique for local thermal annealing is CO_2 laser selective melting [26–29], where an infrared laser source is used to locally melt the surface of an exposed or shallow-buried microchannel. Alternatively, the surfaces can be smoothed with isothermal annealing of the whole volume [30] in a furnace, with a maximum temperature just above the annealing point of the material in order to release the internal stress and to smooth superficial asperities. This approach has the advantage of being suitable for the smoothing of buried microstructures, regardless of their orientation or geometry, and is particularly appealing for optofluidic applications due to the possibility to improve the surface quality without altering the surface profile.

Starting from these previous results, in this work, we present a comprehensive study that, for the first time, combines FLM irradiation parameter optimization and thermal annealing, demonstrating the possibility to highly improve the surface quality of optical elements. We realized flat surfaces with the FLICE technique and HF etching and characterized their roughness in terms of average values and spatial features in order to optimize the laser-writing geometry. Afterwards, we define a thermal annealing procedure for surface smoothing and quantify its effects with comparative analysis. Lastly, as a proof of concept, we show the effectiveness of thermal smoothing with the realization of millimetric embedded mirrors. The analysis of reflected images acquired through a treated and untreated mirror highlights a clear surface quality improvement after the thermal annealing.

2. Results

2.1. Optimization of Laser Irradiation

In our optimization work, we decided to start from the most simple 2D surface, i.e., a flat surface, written along the laser propagation axis (z axis—see Section 4.1). Because of mounting constraints of the sample in the measurement instrumentation, we realized millimetric-sized surfaces on a paralellepiped of 2 mm by 1 mm by 1 mm.

The first step in the analysis consisted of studying the effect of slot irradiation geometry (see Section 4.1 for further details) on the final roughness. All laser irradiation parameters were chosen from previous experiments in order to guarantee a uniform etching speed at different sample depths. We realized 5 different samples by varying the slot thickness, the vertical pitch between the irradiation lines, and the orientation with respect to the laser polarization. All details are reported in Table 1.

Table 1. Sample fabrication parameters and measured root mean squared (RMS) roughness before thermal annealing (S_q). *Pol* indicates the polarization orientation with respect to the line writing direction, and *dz* is the spacing between horizontal planes.

Sample	N_{lines}	dz	Pol	S_q
A	6 lines	2 μm	⊥	43 nm
B	single line	2 μm	⊥	57 nm
C	6 lines	5 μm	⊥	28 nm
D	6 lines	10 μm	⊥	35 nm
E	6 lines	2 μm	‖	63 nm

The measured root mean squared (RMS or S_q) roughnesses are reported in the table. The results are in line with the literature, and it is confirmed that a polarization orthogonal with respect to the sample translation direction gives a lower S_q value. Interesting considerations could be inferred when looking at the spatial frequency components of the profiles, as evident in Figure 1, where the 2D power spectral densities (PSDs^{2D}) of three samples are shown. The data of sample A present a radially symmetrical distribution, representing a randomly distributed height profile. On the other hand, sample D, written with a coarser spacing of the laser lines along z direction, shows a sharp peak at the spatial frequency corresponding to a spatial wavelength of 10 μm. It has to be underlined that these peaks correspond to the sample z spacing; thus, it can be linked to periodic modulation of the surface as reminiscence of the irradiation pattern. A similar effect was seen also in sample E (not reported) even with a lower contrast. A secondary peak, corresponding to a spatial wavelength of 50 μm, is also visible, but it is present also in all samples and it has been attributed to the Fourier Transform windowing effect. Lastly, sample E, written parallel to the laser polarization, presents a random distribution profile, coherent with the lack of a preferential etching direction and thus with a more homogeneous acid attack [21].

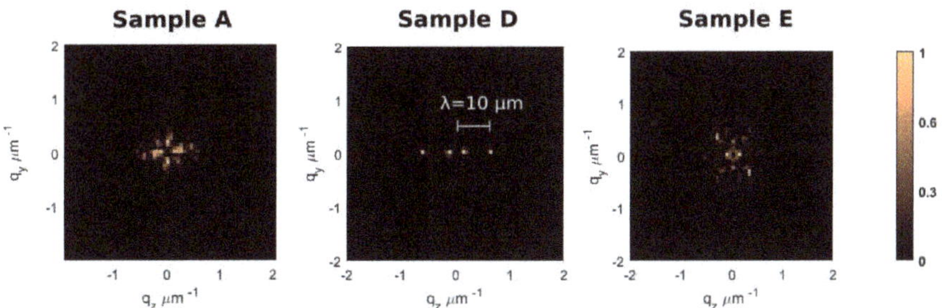

Figure 1. Two-dimensional power spectral densities (PSD) of three samples before thermal annealing. Sample A was written with standard parameters. Sample D had the coarser z spacing (dz). Sample E was written with standard parameters but with parallel linear polarization. Each plot is normalized with respect to its maximum.

Analysis of the spatial distribution of the roughness suggests that the z spacing is fundamental in order to control surface characteristics. Indeed, a too coarse pitch may lead to a non-uniform superposition of laser tracks along z direction, even if the spacing is smaller than the dimension of the modified material spot (a typical z dimension of a laser track corresponds to 20–30 μm depending on the depth in the sample and on the focusing objective). As a consequence, a periodic texturing arises along the z direction. This effect can be explained by looking at the evolution of the etching track during time. After the removal of the irradiated volume, the single laser line enlarges isotropically around its axis, until it coalesces with the adjacent tracks in a more random geometry. This type of phenomena is known in lithography, studying the evolution of surface scratches when exposed to HF [31]. Maintaining the etching time constant, the samples with a finer irradiation pattern develop a more random surface texture with respect to those with a coarser spacing. In order to mitigate the contribution of high-frequency components, one can increase the etching time [31] at the cost of a higher deformation of the original irradiated geometry [7]. In view of these considerations, we chose 5 μm z spacing as our upper limit. Once we guaranteed a sufficient superposition of the irradiation lines, the roughness distribution became more random, with a PSD^{2D} profile similar to a single pole exponential decay, typical of mechanically grinded surfaces [32].

2.2. Thermal Annealing Characterization

The glass samples were treated with isothermal annealing with the recipe described in Section 4. The procedure was optimized to reduce the possibility of generation of cracks on the exposed surface. Several test samples were inspected at the optical microscope and scanning electron microscopy (SEM) in order to qualitatively evaluate the surface quality and to optimize the annealing procedure. During this optimization, we used test samples with a parallelepiped shape, such as the one presented before, and FLICE-microstructured samples with surface-exposed structures such as wedges or cones. An example of these is reported in Figure 2. In subfigures (a) and (b), the SEM image of a cone with a central circular microchannel is reported before and after the thermal annealing. The reduction in surface roughness is qualitatively clear.

We studied the effects of the treatment on three samples, samples B and C, that present higher and lower roughnesses for perpendicular polarization, and sample E, written with parallel polarization. The three samples were analyzed a second time with the stylus profilometer, focusing on an area in the same position as the previous measurement. We obtained a reduced S_q in all three cases, 25 nm, 20 nm, and 12 nm, for samples B, C, and E, respectively, starting from values of 57 nm, 28 nm, and 63 nm. In Figure 2c, the 1D PSD of sample C is shown, obtained by averaging the PSD^{2D} along y axis. It is clear that the annealing reduced the RMS roughness (that corresponds to the area under the graph) and that it had a higher influence on the high-frequency components, i.e., on the right-handed part of the plot. This is coherent with the qualitative model presented for other thermal treatments [17], stating that only the first layers of the surface undergo softening or partial melting, whereas the form, i.e., the low-frequency components, is less affected. This guarantees better preservation of the microstructured profile during the smoothing process.

These results show that isothermal annealing can actually improve the surface roughness of surfaces of embedded optical elements, smoothing those frequency components that are responsible for undesired scattering and that cause minor modifications on the form of the element.

As a proof of concept of the efficiency of the thermal smoothing, we realized and tested an embedded flat mirror. This mirror consists of hollow slot along the z axis realized with the same irradiation geometry shown before. In order to appreciate the removal of surface patterning after the smoothing, we chose to use the sample C irradiation parameters. The slot was oriented at 45° with respect to the glass sample facets and was used as a total internal reflection-based mirror. The sample was characterized before and after thermal annealing in order to evaluate the changes in the optical performances.

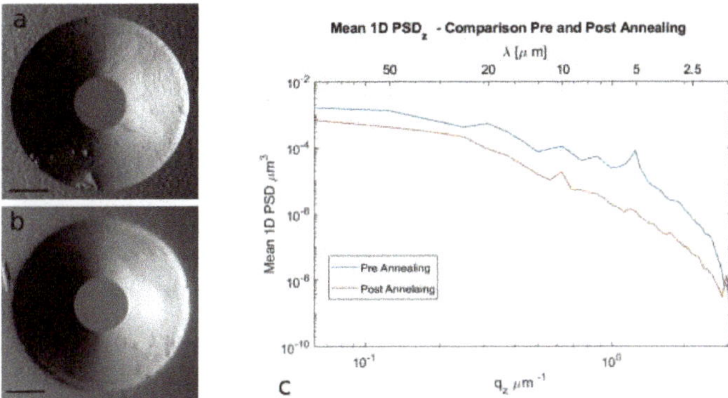

Figure 2. Thermal annealing effect on surface roughness. SEM images of two microstructured cones before (**a**) and after (**b**) thermal treatment. Scale bars correspond to 50 µm; (**c**) comparison of sample C 1D PSD$_z$ before and after the thermal treatment, reported in logaritmic scale.

The scheme of this experiment, together with a drawing of the mirror, is reported in Figure 3a. Collimated light was used to illuminate a target. The transmitted light, once reflected by the embedded mirror, was focused by a microscope objective on a camera. An image of the same target, with the same objective and with a standard silver mirror, was taken as a reference. The three images (reference, not-annealed mirror, and annealed mirror) are reported in Figure 3b. In the case of the reference optical system, we were able to clearly distinguish up to the smallest feature size of the target in analysis (element 7.6 of the USAF-1951 target, corresponding to 2.19 µm). As a consequence, any degradation of the images is imputable to the surface quality of the mirrors. In order to determine the maximum resolution, we used Raylegh criterion: the difference in intensity between two peaks and the dip in between should be greater than 20% of the peak. In the case of the not-treated mirror, we obtained a vertical resolution of 6.96 µm and a horizontal one of 6.20 µm. In the case of the annealed one, we obtained an improved result of 4.38 µm along the vertical direction and 4.92 µm along the horizontal.

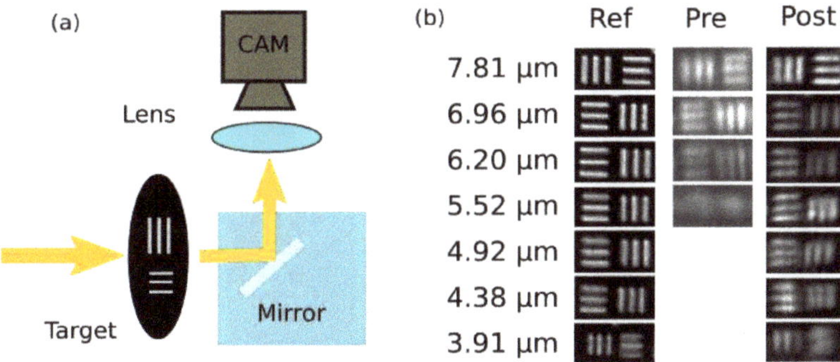

Figure 3. Optical performances comparison: (**a**) scheme of the experiment; (**b**) comparison of USAF-1951 target images, between the reference (Ref), not-treated mirror (Pre), and smoothed mirror (Post). A blurring effect is evident, induced by the residual roughness in the images from not-treated mirrors that contribute to increasing the background and to reducing the contrast.

These results show how the roughness reduction after thermal annealing could be beneficial for the performances of embedded total internal reflection mirrors. The reduction in high-frequency components of the PSD reduces the undesired distortion introduced by the mirror. The difference in resolution between horizontal and vertical directions, i.e., y and z axes of the mirror surface, is coherent with the results on the PSD^{2D}.

3. Discussion

In this work, we analyzed the effect of irradiation geometry optimization and isothermal annealing on the surface quality of micro-structures realized with FLM and HF etching.

First, we verified the results reported in the literature, $S_q = 60 - 30$ nm, over an area of 100 μm by 100 μm per sample. Analysis of the PSD^{2D} shows that increasing the spacing between adjacent irradiation lines could cause periodic patterning on the surface. In our case, a distance grater or equal to 5 μm causes high-frequency peaks in the PSD^{2D}. If the spacing between irradiation lines is sufficiently fine, the PSD^{2D} presents a random-like distribution, both in the case of parallel and perpendicular laser polarizations.

We quantitatively measured the effect of isothermal annealing on the fused silica etched surfaces, characterizing the spatial components that contribute to roughness. Thanks to this analysis, we verified the improvements in the surface quality and the beneficial effect on the high-frequency components of the PSD. The resulting surfaces present an S_q ranging from 25 μm to 12 μm. We verified this improvement by analyzing the optical performances of two flat embedded TIR mirrors, with and without thermal annealing. We showed that the untreated mirrors provide a resolution around 6–7 μm, depending on the axis of analysis. An element such as this could still be used as a low-quality mirror or as spatial filter, but it is not suitable for most photonic applications. On the other hand, we verified that including the smoothed mirror inside an imaging system guarantees a resolution of around 4.5 μm, sufficient for the detection of cell-like samples in biological applications, with a typical dimension in the 5–15 μm range.

This analysis gives useful indications for the design of micro-optical elements in FLICE fabrication. Further analysis could include the characterization of sub-micrometric roughness using AFM or an optical profilometer and evaluation of the thermal annealing effect on more advanced optical elements, such as embedded lenses.

4. Materials and Methods

4.1. FLM Fabrication: Setup and Parameters

The fs-laser source used for laser writing consisted of a commercial Ytterbium-based laser (Spirit 1040-16, Spectra-Physics, Stahnsdorf, Germany) emitting at 1042 nm working at 1 MHz repetition rate and with a pulse duration of approximately ≈400 fs. An external LBO crystal was used for second harmonic (SH) generation, obtaining a green light at 521 nm. The laser SH was focused from the top of the sample using a 50 × 0.6 NA objective (Zeiss, Jena, Germany). The linear laser polarization direction can be adjusted using a half-wave plate placed in the SH generation stage. The sample was mounted on a three-axes motorized stage (FIBERGlide 3D, Aereotech Inc., Pittsburgh, PA, USA). We realized all samples in JG1 fused silica windows of 1 mm thickness (Focktech Photonics, Inc., Fujian, China). The surfaces for the roughness analysis were realized by irradiating 4 slots along the Z direction, i.e., the laser propagation axis (see Figure 4a) in order to realize a parallelepiped that could be then detached from the bulk and separately analyzed. The irradiation pattern of the single surface, shown in the inset of Figure 4a, consisted in several XY planes stacked along the Z direction. Each plane was composed of a variable number of laser lines N_{lines} (from 1 to 6) spaced $d = 1$ μm each. Both the vertical spacing of the planes (dz) and the number of lines per plane were changed from sample to sample to study its impact on the residual roughness. The surfaces had a height of 1 mm (corresponding to the glass thickness) and a width of 2 mm. We used a fixed translation speed of 1.5 mm/s and a fixed pulse energy of 210 nJ, parameters chosen to guarantee nanograting formation at all depths in the sample. The chemical etching was performed using a hydrofluoric acid and deionized water solution at 20% volume concentration. The acid was kept at a constant temperature of 35 °C and we used an ultrasound bath to favor acid diffusion inside the etched slots. The samples were immersed in acid for approximately 1 h.

The embedded mirrors were realized with the same procedure as that of the single surfaces. Moreover, the external perimeter of the device was realized at the same time in order to guarantee perfect alignment between the mirror and the device facets.

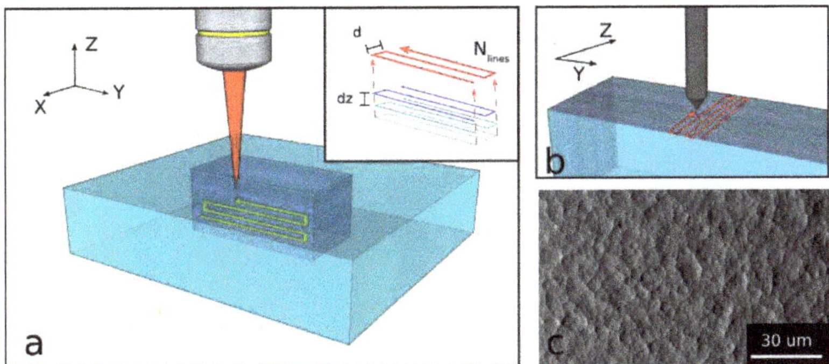

Figure 4. Schemes of the surface fabrication and analysis: (**a**) laser irradiation pattern geometry; (**b**) stylus profilometer raster scan analysis; (**c**) SEM image of a FLICE (femtosecond laser irradiation followed by chemical etching) machined surface.

4.2. Thermal Smoothing

For thermal isotropic annealing of the surfaces, we used a furnace (Nabertherm GmbH, Lilienthal, Germany L5/13/P330) with no controlled atmosphere. Before the treatment, the samples were placed overnight in a solution of sulphuric acid and potassium dichromate and then rinsed in deinonized water to remove surface contamination. The glass parallelepipeds were placed in the center of the furnace, with the surface of interest placed on the top, to avoid contamination from contact with the furnace floor. The procedure consisted of a first heating of the sample up to 1215 °C, with a step at 800 °C for 10 h. The furnace was maintained at maximum temperature for 25 h, before slowly cooling it down to 1000 °C. The sample was then cooled down to room temperature with a fast ramp of −100 °C/h. The details are reported in Figure 5.

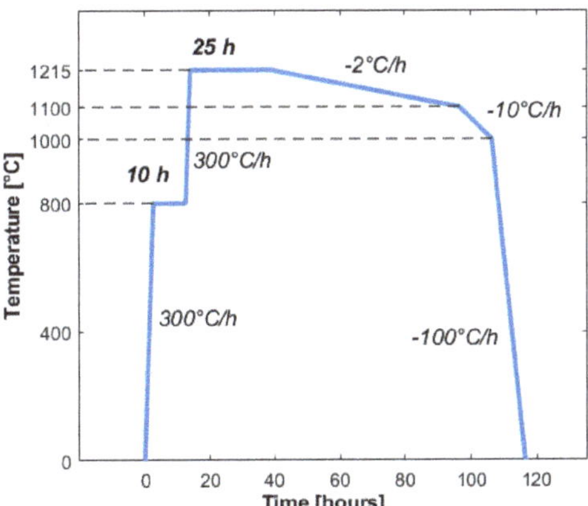

Figure 5. Thermal annealing procedure parameters.

4.3. Roughness Analysis

The treated surfaces have been firstly observed with brightfield transmission microscope and SEM (Phenomenon Pro, Thermo Fisher Scientific Inc., Waltham, MA, USA), looking for major surface defects or the presence of cracks. Then a quantitative analysis was performed using a stylus profilometer (Tencor P-17, KLA Corporation, Milpitas, CA, USA). The instrument allowed us to inspect a surface of 100 µm by 100 µm in approximately 25 min, with a lateral resolution of 1 µm and a minimum detectable Z displacement of 0.2 nm. For consistency between measures, we always analyzed a squared area at 200 µm distance from the inferior facet of the parallelepiped. The measurement was carried out in raster scan mode, i.e., moving the stylus in contact with the surface along parallel lines, as reported in the scheme in Figure 4b. A typical SEM-acquired image is reported in Figure 4c, showing the texture of the surface under analysis. Lo Turco et al. [21] showed how FLICE machined surface characteristics depend not only on the roughness average value but also on its spatial distribution. For this reason, in our study, we focused on two main parameters: the root mean square (S_q or RMS) roughness and the 2D power spectral density (PSD^{2D}). The first is a standard quantification of the average value of the surface height profile, and it is also the standard deviation of the heights. It can be computed as follows:

$$S_q = \sqrt{\frac{1}{A} \iint_A Z^2(x,y) \quad dxdy} \tag{1}$$

where A is the total analyzed area in the xy plane and Z(x,y) is the surface heights profile [32]. The PSD^{2D} provides a representation of the surface roughness as a function of its spatial frequencies, i.e., the inverse of the spatial wavelengths. It can be computed as

$$PSD^{2D}(q_x, q_y) = \frac{1}{A} \quad \left|\tilde{Z}(q_x, q_y)\right|^2 \tag{2}$$

where $\tilde{Z}$ is the discrete Fourier transform of the discrete set of data Z(x,y), and q_x and q_y are the spatial frequencies defined as $q_i = 2\pi * L_i / N_i$, with N_i number of equispaced acquired point in the i direction and with L_i total length along the i direction. The PSD^{2D} of a finite and non-periodic set of data was computed including a Welch window function to smooth the frequency contribution given by the edges of the dataset.

The data were acquired and filtered using the APEX 3D software (KLA Corporation, Milpitas, CA, USA) form-removal function in order to virtually remove the form profile of the surface and to limit the analysis to the proper roughness. All analyses were carried out using custom-made MATLAB (MathWorks) codes. The effects of surface roughness were characterized by looking at the performances of embedded mirrors. The outer edges of the samples were mechanically polished in order to minimize their impact on the light transmission. A USAF-1951 target (Thorlabs GmbH, 85232 Bergkirchen, Germany) was shone with collimated light coming from a microscope lamp. The obtained image was then reflected by the mirrors and projected by a microscope objective (10× 0.25 NA) onto a CMOS camera (Edmund Optics, Barrington, NJ, USA). The acquired images were then analyzed using Fiji (ImageJ, NIH, Bethesda, MD, USA) application to determine the maximum level of resolution achievable.

Author Contributions: Conceptualization R.O., F.B. and P.P.; investigation, F.S., P.P. and R.M.V.; writing—original draft preparation, F.S., P.P. and F.B.; project administration, R.O. and F.B. All authors contributed to editing and revising the text. All authors have read and agreed to the published version of the manuscript.

Funding: This research was partially funded by the European Union under the Horizon 2020 Framework program, grant agreement no.801336 (PROCHIP).

Institutional Review Board Statement: Not applicable.

Informed Consent Statement: Not applicable.

Data Availability Statement: All data needed to support the conclusions of the paper are present in the text. Raw data are available upon request.

Acknowledgments: The experimental characterization was partially performed at PoliFAB, the micro- and nanofabrication facility of Politecnico di Milano (www.polifab.polimi.it (accessed on 7 February 2021)). The authors would like to thank the PoliFAB staff for the valuable technical support.

Conflicts of Interest: The authors declare no conflict of interest.

References

1. Della Valle, G.; Osellame, R.; Laporta, P. Micromachining of photonic devices by femtosecond laser pulses. *J. Opt. A Pure Appl. Opt.* **2009**, *11*, 013001. [CrossRef]
2. Shimotsuma, Y.; Kazansky, P.G.; Qiu, J.; Hirao, K. Self-organized nanogratings in glass irradiated by ultrashort light pulses. *Phys. Rev. Lett.* **2003**, *91*, 1–4. [CrossRef]
3. Richter, S.; Miese, C.; Döring, S.; Zimmermann, F.; Withford, M.J.; Tünnermann, A.; Nolte, S. Laser induced nanogratings beyond fused silica—Periodic nanostructures in borosilicate glasses and ULE. *Opt. Mater. Express* **2013**, *3*, 1161. [CrossRef]
4. Zhang, F.; Zhang, H.; Dong, G.; Qiu, J. Embedded nanogratings in germanium dioxide glass induced by femtosecond laser direct writing. *J. Opt. Soc. Am. B* **2014**, *31*, 860. [CrossRef]
5. Zimmermann, F.; Lancry, M.; Plech, A.; Richter, S.; Hari Babu, B.; Poumellec, B.; Tünnermann, A.; Nolte, S. Femtosecond laser written nanostructures in Ge-doped glasses. *Opt. Lett.* **2016**, *41*, 1161. [CrossRef]
6. Fedotov, S.S.; Drevinskas, R.; Lotarev, S.V.; Lipatiev, A.S.; Beresna, M.; Čerkauskaitė, A.; Sigaev, V.N.; Kazansky, P.G. Direct writing of birefringent elements by ultrafast laser nanostructuring in multicomponent glass. *Appl. Phys. Lett.* **2016**, *108*, 071905. [CrossRef]
7. Vishnubhatla, K.C.; Bellini, N.; Ramponi, R.; Cerullo, G.; Osellame, R. Shape control of microchannels fabricated in fused silica by femtosecond laser irradiation and chemical etching. *Opt. Express* **2009**, *17*, 8685–8695. [CrossRef]
8. Rajesh, S.; Bellouard, Y. Towards fast femtosecond laser micromachining of glass, effect of deposited energy. In Proceedings of the Conference on Lasers and Electro-Optics, San Jose, CA, USA, 16–21 May 2010; p. JTuD18. [CrossRef]
9. Bellouard, Y.; Said, A.A.; Bado, P. Integrating optics and micro-mechanics in a single substrate: A step toward monolithic integration in fused silica. *Opt. Express* **2005**, *13*, 6635. [CrossRef]
10. Osellame, R.; Hoekstra, H.J.; Cerullo, G.; Pollnau, M. Femtosecond laser microstructuring: An enabling tool for optofluidic lab-on-chips. *Laser Photonics Rev.* **2011**, *5*, 442–463. [CrossRef]
11. Sala, F.; Castriotta, M.; Paiè, P.; Farina, A.; D'Annunzio, S.; Zippo, A.; Osellame, R.; Bragheri, F.; Bassi, A.; Osellame, R.; et al. High-throughput 3D imaging of single cells with light-sheet fluorescence microscopy on chip. *Biomed. Opt. Express* **2020**, *11*, 4397. [CrossRef]
12. Paiè, P.; Zandrini, T.; Vázquez, R.M.; Osellame, R.; Bragheri, F. Particle manipulation by optical forces in microfluidic devices. *Micromachines* **2018**, *9*, 200. [CrossRef]
13. Cheng, Y.; Sugioka, K.; Midorikawa, K.; Masuda, M.; Toyoda, K.; Kawachi, M.; Shihoyama, K. Three-dimensional micro-optical components embedded in photosensitive glass by a femtosecond laser. *Opt. Lett.* **2003**, *28*, 1144. [CrossRef] [PubMed]
14. Cheng, Y.; Sugioka, K.; Midorikawa, K. Microfluidic laser embedded in glass by three-dimensional femtosecond laser microprocessing. *Opt. Lett.* **2004**, *29*, 2007–2009. [CrossRef]
15. Wang, Z.; Sugioka, K.; Midorikawa, K. Three-dimensional integration of microoptical components buried inside photosensitive glass by femtosecond laser direct writing. *Appl. Phys. A Mater. Sci. Process.* **2007**, *89*, 951–955. [CrossRef]
16. Simoni, F.; Bonfadini, S.; Spegni, P.; Lo Turco, S.; Lucchetta, D.; Criante, L. Low threshold Fabry-Perot optofluidic resonator fabricated by femtosecond laser micromachining. *Opt. Express* **2016**, *24*, 17416. [CrossRef]
17. He, F.; Cheng, Y.; Qiao, L.; Wang, C.; Xu, Z.; Sugioka, K.; Midorikawa, K.; Wu, J. Two-photon fluorescence excitation with a microlens fabricated on the fused silica chip by femtosecond laser micromachining. *Appl. Phys. Lett.* **2010**, *96*, 041108. [CrossRef]
18. Paiè, P.; Bragheri, F.; Claude, T.; Osellame, R. Optofluidic light modulator integrated in lab-on-a-chip. *Opt. Express* **2017**, *25*, 7313. [CrossRef]
19. Watanabe, W.; Kuroda, D.; Itoh, K.; Nishii, J. Fabrication of Fresnel zone plate embedded in silica glass by femtosecond laser pulses. *Opt. Express* **2002**, *10*, 978. [CrossRef] [PubMed]
20. Nieto-Vesperinas, M.; Sánchez-Gil, J.A. Light scattering from a random rough interface with total internal reflection. *J. Opt. Soc. Am. A* **2008**, *9*, 424. [CrossRef]
21. Lo Turco, S.; Di Donato, A.; Criante, L. Scattering effects of glass-embedded microstructures by roughness controlled fs-laser micromachining. *J. Micromech. Microeng.* **2017**, *27*, 65007. [CrossRef]
22. Dogan, Y.; Madsen, C.K. Optimization of ultrafast laser parameters for 3D micromachining of fused silica. *Opt. Laser Technol.* **2019**, *123*, 105933. [CrossRef]
23. Ho, S.; Herman, P.R.; Aitchison, J.S. Single- and multi-scan femtosecond laser writing for selective chemical etching of cross section patternable glass micro-channels. *Appl. Phys. A Mater. Sci. Process.* **2012**. [CrossRef]
24. Ross, C.; Maclachlan, D.G.; Choudhury, D.; Thomson, R.R. Towards optical quality micro-optic fabrication by direct laser writing and chemical etching. *Front. Ultrafast Opt. Biomed. Sci. Ind. Appl. XVII* **2017**, *10094*, 100940V. [CrossRef]
25. He, F.; Cheng, Y.; Xu, Z.; Liao, Y.; Xu, J.; Sun, H.; Wang, C.; Zhou, Z.; Sugioka, K.; Midorikawa, K.; et al. Direct fabrication of homogeneous microfluidic channels embedded in fused silica using a femtosecond laser. *Opt. Lett.* **2010**, *35*, 282–284. [CrossRef]
26. Lin, J.; Yu, S.; Ma, Y.; Fang, W.; He, F.; Qiao, L.; Tong, L.; Cheng, Y.; Xu, Z. On-chip three-dimensional high-Q microcavities fabricated by femtosecond laser direct writing. *Opt. Express* **2012**, *20*, 10212. [CrossRef] [PubMed]
27. Jung, S.; Lee, P.A.; Kim, B.H. Surface polishing of quartz-based microfluidic channels using CO_2 laser. *Microfluid. Nanofluid.* **2016**, *20*, 84. [CrossRef]

28. Serhatlioglu, M.; Ortaç, B.; Elbuken, C.; Biyikli, N.; Solmaz, M.E. CO_2 laser polishing of microfluidic channels fabricated by femtosecond laser assisted carving. *J. Micromech. Microeng.* **2016**, *26*. [CrossRef]
29. Weingarten, C.; Steenhusen, S.; Hermans, M.; Willenborg, E.; Schleifenbaum, J.H. Laser polishing and 2PP structuring of inside microfluidic channels in fused silica. *Microfluid. Nanofluid.* **2017**, *21*, 165. [CrossRef]
30. He, F.; Lin, J.; Cheng, Y. Fabrication of hollow optical waveguides in fused silica by three-dimensional femtosecond laser micromachining. *Appl. Phys. B Lasers Opt.* **2011**, *105*, 379–384. [CrossRef]
31. Feit, M.D.; Suratwala, T.I.; Wong, L.L.; Steele, W.A.; Miller, P.E.; Bude, J.D. Modeling wet chemical etching of surface flaws on fused silica. *Proc. SPIE* **2009**, *7504*, 75040L. [CrossRef]
32. Bhushan, B. Surface roughness analysis and measurement techniques. *Mod. Tribol. Handb. Vol. One Princ. Tribol.* **2000**, *1*, 49–119. [CrossRef]

MDPI

Review

Femtosecond-Laser Assisted Surgery of the Eye: Overview and Impact of the Low-Energy Concept

Catharina Latz [1,*], Thomas Asshauer [2], Christian Rathjen [2] and Alireza Mirshahi [1]

[1] Dardenne Eye Hospital, D-53173 Bonn, Germany; mirshahi@dardenne.de
[2] Ziemer Ophthalmic Systems, CH-2562 Port, Switzerland; Thomas.Asshauer@ziemergroup.com (T.A.); christian.rathjen@ziemergroup.com (C.R.)
* Correspondence: latz@dardenne.de

Abstract: This article provides an overview of both established and innovative applications of femtosecond (fs)-laser-assisted surgical techniques in ophthalmology. Fs-laser technology is unique because it allows cutting tissue at very high precision inside the eye. Fs lasers are mainly used for surgery of the human cornea and lens. New areas of application in ophthalmology are on the horizon. The latest improvement is the high pulse frequency, low-energy concept; by enlarging the numerical aperture of the focusing optics, the pulse energy threshold for optical breakdown decreases, and cutting with practically no side effects is enabled.

Keywords: femtosecond laser; fs-assisted cataract surgery; laser-assisted ophthalmic surgery; high pulse frequency; low energy

Citation: Latz, C.; Asshauer, T.; Rathjen, C.; Mirshahi, A. Femtosecond-Laser Assisted Surgery of the Eye: Overview and Impact of the Low-Energy Concept. *Micromachines* **2021**, *12*, 122. https://doi.org/10.3390/mi12020122

Academic Editor: Rebeca Martínez Vázquez

Received: 15 December 2020
Accepted: 18 January 2021
Published: 24 January 2021

1. Introduction

Laser technology and ophthalmic surgery have shaped each other over the past 40 years. The optically transparent structures of the eye, namely cornea, lens, and vitreous body, allow for delivery of the laser energy at different focal depths, thereby giving access to surgical interventions without having to open or mechanically enter the eye (Figure 1). Other types of lasers, with various wavelengths, pulse durations, and power levels, interact with eye tissues in a range of different ways. For continuous laser irradiation of low to moderate average power (mW range), photochemical and thermal effects induced by the absorbed light are the dominant laser–tissue interactions. Depending on the wavelengths used, specific types of molecules can be optically excited to trigger chemical reactions, or local heating of specific tissue can be achieved. If temperatures above 60 °C are reached, tissue coagulation will occur. When pulsed laser light with intensities between 10^7 and 10^9 W/cm^2 interacts with strongly absorbing tissue, near-surface material can be removed explosively. This effect is called "photoablation". In ophthalmology, it is applied to change the curvature of the cornea with pulsed UV light from excimer lasers. For shorter pulse durations in the ps to fs range and even higher intensities above 10^{11} W/cm^2, more exotic interactions can be achieved, as will be explained in detail below. A more comprehensive general overview of laser–tissue interaction mechanisms can be found in excellent quality in several text books [1,2].

The first reported ophthalmic use of short pulse lasers at near-infrared wavelengths was in 1979 by Aron-Rosa, who treated posterior capsule opacification (PCO) after cataract surgery [3]. In 1989, Stern et al. demonstrated that by decreasing pulse width of ultrashort-pulsed lasers from nano- to femtoseconds (ns: 10^{-9} s, fs: 10^{-15} s), ablation profiles showed higher precision and less collateral damage [4]. At the same time, optical coherence tomography developed and provided noninvasive three-dimensional (3D) in vivo imaging with fine resolution in both lateral and axial dimensions at a micrometer level [5]. These developments offered ophthalmic surgeons a tool for high precision cutting and visual control through imaging, and ultimately allowed a gamut of treatment applications for these lasers within the field of ophthalmology. Recent changes in the numerical aperture of the laser focusing optics and the repetition rate of the laser sources have further decreased collateral damage while increasing precision. This review article gives an overview of the technical backgrounds of femtosecond lasers and OCT imaging as well as clinical applications in ophthalmic surgery today.

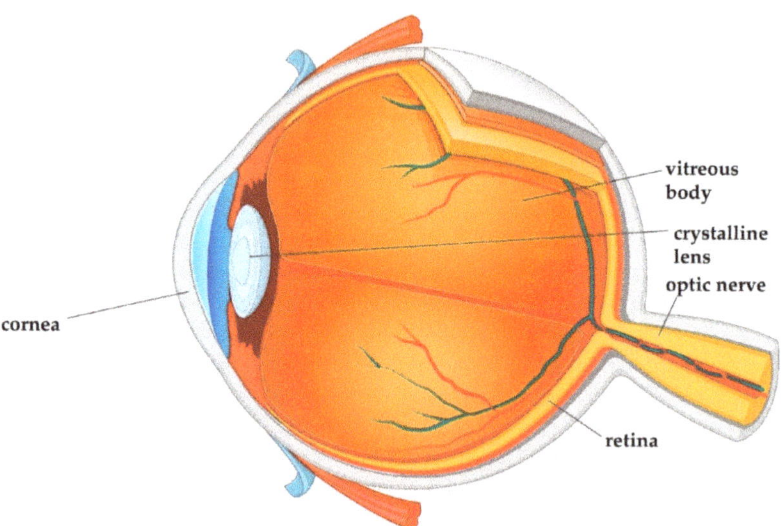

Figure 1. Cross-section of the eye. Cornea, crystalline lens, and vitreous body are transparent in the healthy eye.

2. Laser Technology

2.1. Solid-State Lasers in Ophthalmology

2.1.1. Nd:YAG Laser with Ns Pulse Durations

The first type of short-pulsed laser at near-infrared wavelengths successfully used in ophthalmology was the Q-switched Nd:YAG solid-state laser. Its wavelength of 1064 nm is transmitted by all the visually transparent structures in the eye (cornea, lens, and vitreous body). Their pulse durations are a few nanoseconds (ns), and for ophthalmic applications, pulse energies in the range of 0.3–10 mJ are typically used [6].

When Nd:YAG laser pulses are strongly focused at a location inside the eye, to spot sizes in the order of a few microns, the combination of short pulse duration focusing to minimal spot sizes creates very high intensities at the laser focus, above 10^{11} W/cm^2. Under these conditions, a phenomenon called "optical breakdown" occurs. In the first step, multiphoton absorption leads to ionization of some tissue molecules, creating free electrons. In the subsequent second step, these "seed" electrons absorb photon energy and are thus accelerated. After repeated photon absorptions, electrons reach a sufficiently high kinetic energy to ionize themselves more molecules by impact ionization, creating more free electrons. If the laser irradiation is intense enough to overcome electron losses, an avalanche effect occurs [2].

When the extremely fast rising electron density exceeds values of approximately 10^{20}/cm^3, a "plasma state of matter" (cloud of ions and free electrons) is created at the laser focus [2]. This plasma is highly absorbing for photons of all wavelengths. Therefore, the rest of the laser pulse is directly absorbed by the plasma, increasing its temperature and energy density (Figure 2).

The hot plasma cloud rapidly recombines to a hot gas, with a thermalization time of the energy initially carried by free electrons of a few picoseconds to tens of ps [7]. This time is much shorter than the acoustic transit time from the center of the focus to the periphery of the plasma volume, leading to confinement of the thermoelastic stresses caused by the temperature rise. Conservation of momentum requires that the stress wave emitted in this geometrical configuration contains both compressive and tensile components [7]. If sufficient pulse energy density is applied, the tensile stress wave becomes strong enough to induce fracture of the tissue, causing the formation of a cavitation bubble [7]. Depending on the pulse energy, the pressure wave can reach supersonic speed a (shock wave). The high plasma temperature also leads to almost immediate evaporation of the tissue within the focal volume, generating water vapor and gases such as H_2, O_2, methane, and ethane [8]. The resulting gas pressure pushes the surrounding tissue further away, adding to the expansion of the short-lived bubble inside the tissue (Figure 2). The maximum volume temporarily achieved by the bubble scales with the pulse energy above the threshold for optical breakdown. During bubble expansion, the inside pressure ultimately drops below atmospheric pressure due to the outward moving material's inertia, resulting in the bubble dynamically collapsing. The bubble collapse may create another shock wave [2]. This combined process is called "photodisruption" of tissue.

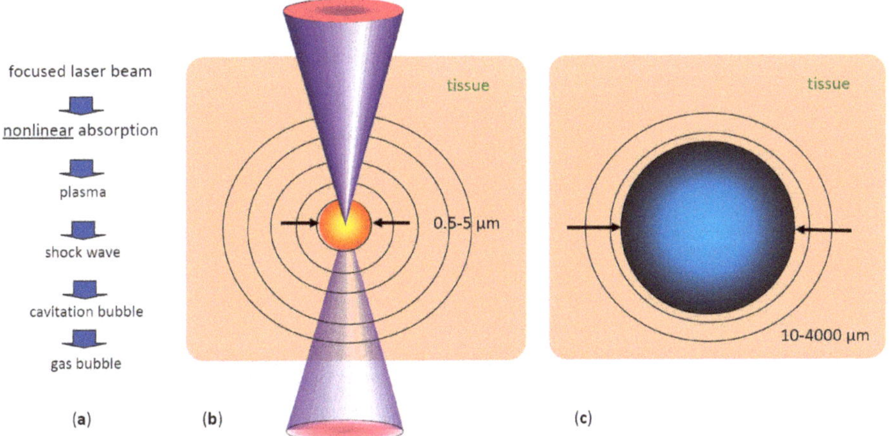

Figure 2. Short pulse laser effects in tissue: (**a**) sequence of effects and induced events, (**b**) plasma size range and pressure wave pattern, (**c**) range of possible cavitation bubble dimensions (pulse energy-dependent) [9].

With typical ophthalmic Nd:YAG laser pulse energies, cavitation bubble radii are in the range of 1000–2000 µm, and shock wave amplitudes at 1 mm distance from the focus reach 100–500 bar [10]. These rather pronounced mechanical side effects restrict the use of Nd:YAG lasers. When shorter pulse ps (10^{-12} s) lasers became available, their mechanical side effects proved to be still too large for delicate tasks as required for ophthalmic applications. This limits Nd:Yag laser application in today's clinical ophthalmological use to cutting isolated tissues, such as the lens capsule in posterior capsular opacification in pseudophakes or small areas of iris tissue to improve aqueous dynamics within the eye.

2.1.2. Femtosecond Lasers

Femtosecond lasers are a more recent advance in solid-state laser technology. They operate at near-infrared wavelengths similar to Nd:YAG lasers but at pulse durations of less than 1 picosecond (ps). As the threshold radiant exposure (J/cm^2) for inducing optical breakdown in tissue is about two orders of magnitude lower in the fs pulse duration regime than at 10 ns [11], much lower pulse energies can be applied to separate tissue. High pulse repetition rates from 10 s of kHz to even MHz are then used to create continuous cut planes inside the tissue by placing many pulses close to each other with three-dimensional beam scanning systems.

The lower pulse energies lead to a drastic reduction of the mechanical side effects of optical breakdown. For 300 fs pulses of 0.75 μJ energy, the generated cavitation bubbles have radii of only 45 μm, almost two orders of magnitude smaller than ns pulse with energies in the mJ range [12]. In addition, the associated pressure waves are much weaker, 1–5 bar at 1 mm distance [13]. This process is referred to as "plasma-induced ablation", as the disruptive mechanical side effects of ns pulses described above are absent. Additionally, the thermal side effects of fs pulses in tissue are almost negligible [7].

The first commercially available, USA Food and Drug Administration (FDA)-approved fs-laser system for ophthalmology, the IntraLase™ FS, was launched in 2001 [14]. It was used for "flap" creation in LASIK refractive surgery (see Section 3.1.1 below), replacing mechanical cutting devices called microkeratomes. Its first commercial version operated at a 15 kHz repetition rate and pulse energies of several μJ [15]. Further fs-laser systems for "flap" cutting and other corneal surgery were launched by several manufacturers in the following years, including the Ziemer FEMTO LDV in 2005, which introduced a new concept of low pulse energies and high repetition rates, and later the Wavelight FS200 and the Zeiss VisuMax™.

In 2009, the LensX™ system was introduced, the first commercial fs laser designed for cataract surgery, thus opening a new field of fs-laser application within ophthalmology [16]. Its early versions operated at 33 kHz repetition rate and pulse energies of 6–15 μJ [17]. LensX became part of Alcon, and again, in the following years, multiple other manufacturers launched similar products, including the Johnson & Johnson Optimedica Catalys™, the LENSAR® and the Bausch and Lomb Victus™.

2.1.3. Modern Low Pulse Energy High Repetition Rate Fs Lasers

The pulse energy required to achieve optical breakdown can be reduced in two ways:

First, by shortening the pulse duration—the latest fs lasers can achieve pulse durations of 200–300 fs, while earlier models had pulse durations of up to 800 fs.

Second, by reducing the focal spot size—the focal volume of a Gaussian laser beam is dependent on the axial extension, the so-called Rayleigh range ($z_R = \pi w_0^2/\lambda$) and the beam waist $w_0 = f\lambda/\pi w_L$, where f is the focal length of the lens, w_0 the beam radius at the focus, and w_L the beam radius at the focusing lens. In other words, the focal volume varies inversely with the cube of the numerical aperture NA = w_L/f of the focusing optics (Figure 3). The larger the numerical aperture NA, the smaller the focal spot and finally, the smaller the energy threshold for optical breakdown [18].

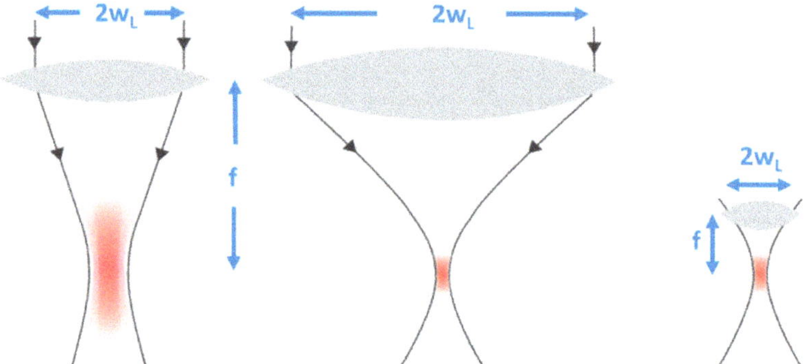

Figure 3. The focal volume of a Gaussian laser beam scales with the numerical aperture NA = w/f of the focusing lens. The larger the NA, the smaller the focal spot volume.

To practically achieve high NA focusing optics, either the lens diameter can be increased, which quickly becomes bulky and expensive, or the focusing optics can be positioned closer to the eye. The latter approach was implemented by Ziemer Ophthalmic

Systems, using a microscope lens with a short focal length as focusing optics and guiding the laser beam via an articulated mirror arm to a handpiece containing the focusing optics, which is docked to the eye at a short distance.

In 2014, the first low pulse energy fs-laser system for cataract and cornea surgery, the Ziemer FEMTO LDV Z8^{TM}, was CE-approved and commercially launched. It was more compact and lightweight than its predecessors, enabling mobile use.

2.2. Femtosecond Laser–Tissue Interaction

Based on the above laser parameters, the nature of the cutting processes of the two groups differs. In the high pulse energy laser group, the cutting process is driven by mechanical forces applied by the expanding bubbles. The bubbles disrupt the tissue at a larger radius than the plasma created at the laser focus (Figure 4a). On the other hand, in the low pulse energy group, spot separations smaller than the spot sizes are used for overlapping plasmas, which directly evaporate the tissue inside the plasma volume, effectively separating tissue without a need for secondary mechanical tearing effects (Figure 4b). Due to the high pulse repetition rates applied (MHz range), the cutting speeds achieved are similar to the high energy laser group.

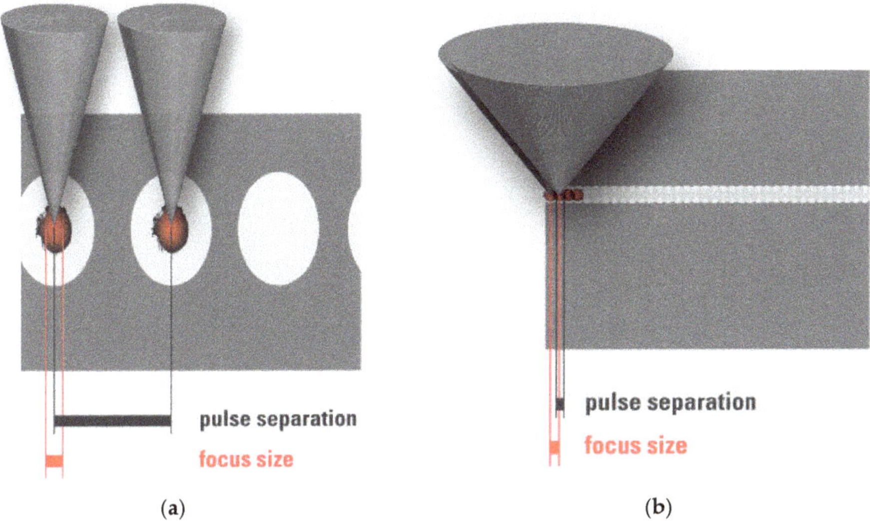

Figure 4. (**a**) High pulse energy, low repetition rate (large spot separation); (**b**) low pulse energy, high repetition rate (small spot separation, overlapping plasma effects of spots).

The cuts achieved by overlapping plasma evaporation of tissue by low energy pulses, however, have a uniquely smooth surface with virtually no damage to the adjacent tissue [19]. This is important for the quality of corneal "flaps", lenticules, or also smooth rims of capsulotomy cuts (see Sections 3.1.1, 3.1.2 and 3.3.2 below). High energy pulses with low repetition rate, on the other hand, rely on the mechanical tearing of tissue in between the actual laser foci. This tearing is accompanied by more stress or potentially even damage to the adjacent tissue [20], as shown by the levels of proinflammatory metabolics detected after laser treatments [21,22].

Software arranges the laser spots in the tissue into geometrical patterns. The software also uses scanning systems to position the laser foci in lines, planes, or even 3D geometries. An example of a 3D cut pattern used for cataract lens fragmentation (see Section 3.3.2 below), which combines multiple planes and cylinders, is shown in Figure 5.

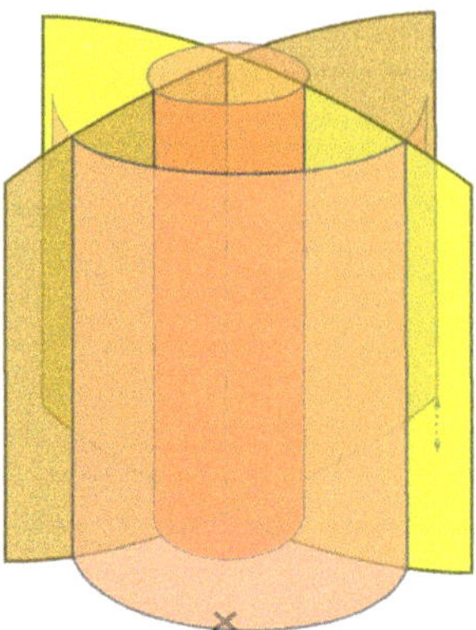

Figure 5. Three-dimensional laser focus scan pattern used for the fragmentation of cataractous lenses.

The energy of fs lasers with wavelengths in the 1030–1060 nm range is transmitted very well through all transparent structures of the eye. However, opaque material scatters the laser radiation and thus reduces the amount of energy reaching the laser focus. For example, laser cutting the cornea at locations with scars requires higher pulse energies than in normal clear cornea. The energy losses depend on the thickness of the scattering material that the laser light is traveling through before reaching the focus. Therefore, the energy loss is more severe when cutting through a several mm thick nucleus of a cataractous lens than through corneal scars, which are only fractions of a mm thick.

The initial fs-laser systems designed for cataract surgery overcame this by using much higher pulse energies than fs lasers for cornea surgery. In the latest generation of versatile multipurpose ophthalmic fs-laser systems, the pulse energy is adaptable over an extensive range, so that for each situation, the adapted amount of pulse energy can be used, but not more, to minimize side effects, such as excessive gas production.

2.3. Supporting Technology Needed in Ophthalmic Fs-Laser Systems

To make an fs-laser device practical for clinical use, some critical supporting technologies needed to be developed as well. Most notable is optical coherence tomography (OCT) imaging of tissue structures, required for the precise positioning of laser cuts deep inside the eye, and the patient interface system using sterile vacuum docking methods to reliably connect the eye to the optical laser delivery system during treatment.

2.3.1. OCT Imaging

OCT is an optical technology that allows for scanning structures inside tissues, thus generating images [23,24]. The images appear similar to ultrasound images but with higher resolution.

The first application of OCT for biological purposes was described by Adolf Fercher et al. for the in vitro measurement of the axial eye length in 1988 (FERCHER 1988). The early clinical OCT systems used so-called time-domain (TD) OCT technology, where the length of the reference arm of an interferometer is mechanically changed. Due to speed limits of this process, these early devices were limited to one-dimensional scans (A-scans), or later small time consuming 2D scans. The so-called frequency-domain OCT (FD-OCT) technology meant a technological breakthrough—it used a fixed reference arm length but a spectrometer with a linear detector array instead of a single detector. Optical path length differences between the interferometer arms in this case produce a periodic modulation in the interference spectrum. By Fourier transformation, an entire A-scan can be retrieved from the measured spectrum [2]. FD-OCT enabled much higher scan speeds, making 2D-imaging and even 3D-imaging feasible in clinical ophthalmology. The first ophthalmic application of FD-OCT, also known as "Fourier domain", was published in 2002 [25].

Later, a further improved variation of frequency-domain OCT technology was developed, "swept-source" (SS) OCT. In this case, a tunable light source with a frequency sweep indicated by a "sawtooth" frequency profile over time is used in combination with a fast single-pixel detector instead of a spectrometer. For further details of OCT technology, and advantages and limitations of its different versions, Section 7.3 of the textbook by Kaschke et al. [2] provides a comprehensive overview and additional literature references.

The initial ophthalmic use of OCT was exclusively for retinal imaging. Starting in 1994, the technology was also developed for imaging the anterior segment of the eye [26]. The possibility of quickly creating high-resolution cross-section images of the cornea, anterior chamber, and lens was a prerequisite for practical cataract surgery laser systems. Imaging and OCT guided surgery was first envisioned by Zeiss and first demonstrated for femtosecond laser surgery by H. Lubatschowski et al. [27].

In most modern cataract fs-laser systems, three-dimensional OCT scans are performed after docking the laser interface to the eye. The LensAR system uses a different technology, a proprietary 3D confocal structured illumination combined with Scheimpflug imaging [28]. In both cases, the resulting images are then analyzed by image processing software, identifying the tissue boundaries of interest [29]. These are notably the anterior and posterior sides of the cornea, the anterior and posterior surfaces of the lens, and the iris (see Figure 6).

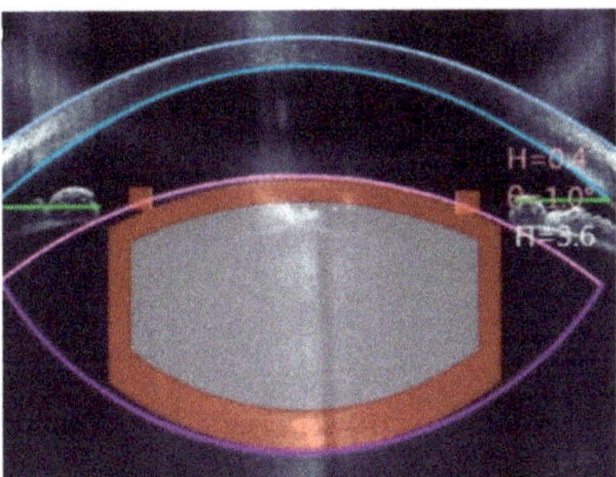

Figure 6. Example of the optical coherence tomography (OCT)-guided placement of an fs-laser cut pattern (blue: corneal anterior and posterior surface; pink and purple: lens anterior and posterior surface; green: iris plane).

This information is used to automatically propose the suitable positions inside the eye for the planned laser cuts, which are also displayed on screens for checking and confirmation by the eye surgeon (Figure 6).

2.3.2. Vacuum Docking Interfaces

For some laser systems, the patient's head is placed under a gantry containing focusing optics at a sufficiently long distance to allow the patient's head to move in and out. In other systems, an articulated arm with a handpiece with focusing optics at its end is used. Due to the flexible arm, the optics can be moved very close to the eye (Figure 7).

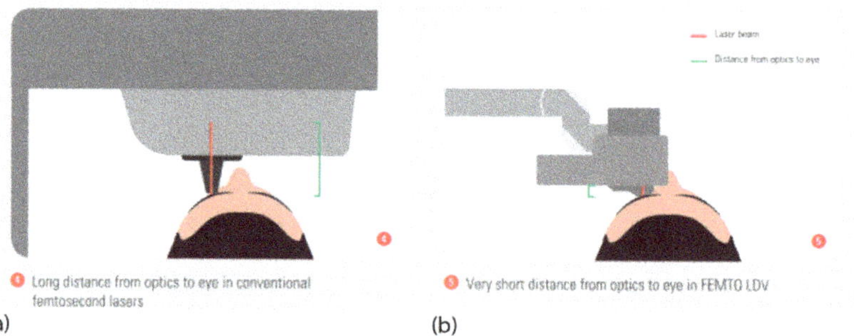

Figure 7. Typical eye docking methods of fs lasers: (**a**) head under fixed laser housing, (**b**) articulated arm with handpiece placed onto the eye; green: distance of eye surface to laser optics.

The eye's actual contact is established via sterile, single-use parts, so-called "patient interfaces". Two different types are in use: (a) applanating interface with a curved or flat interface directly touching the cornea, and (b) liquid-filled interface, where a vacuum ring creates contact to the sclera or the outer cornea, and the center is filled with liquid. The liquid-filled interface allows laser energy transmission while leaving the cornea in its natural shape (Figure 8) [30]. Although contact interfaces temporarily change the shape of the cornea [31], the mechanical contact stabilizes the cornea during surgery to a high degree. This is of particular importance in refractive surgery where precise cuts are required and tissue displacement on a micrometer level has to be avoided. With the absence of clear clinical drawbacks in refractive surgery [32–34], contact interfaces will play a dominant role in the future in corneal surgery. Liquid-filled interfaces with little disturbance of the eye might turn out to be the preferred solution in cataract surgery.

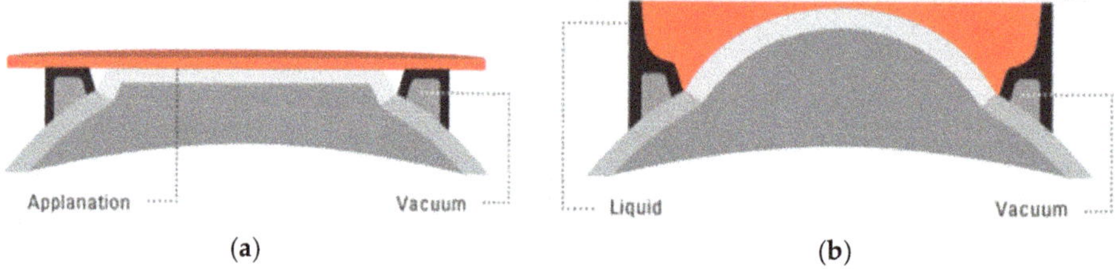

Figure 8. Typical patient interface designs: (**a**) contact interface in direct touch with the cornea (flat or curved), and (**b**) liquid optics interface, no direct touch on the cornea, no deformation.

The stability of the vacuum docking contact during laser emission is of primordial importance. Loss of contact harbors the risk of cutting in wrong planes. Therefore, all lasers are designed to automatically monitor vacuum levels, sometimes complemented with imaging of the eye position (eye tracking), and to immediately stop laser emission upon loss of contact. Of course, the eye surgeons also monitor their patients during the procedure and can manually interrupt or temporarily pause the treatment when they anticipate problems. In case of laser systems with an articulated arm, the surgeons can also use their substantial manual skills to actively stabilize the laser handpiece while in contact with the eye. In any case, after a vacuum loss, the treatment can usually be resumed immediately after a new docking.

3. Clinical Applications

3.1. Refractive Surgery

The human eye functions like the lens of a camera. Images are focused on the retina through a converging system composed mainly of the cornea. If the corneal curvature and thus its refractive power does not precisely match the axial length of the eye, refractive problems like near-sightedness (myopia) or far-sightedness (hyperopia) ensue (Figure 9).

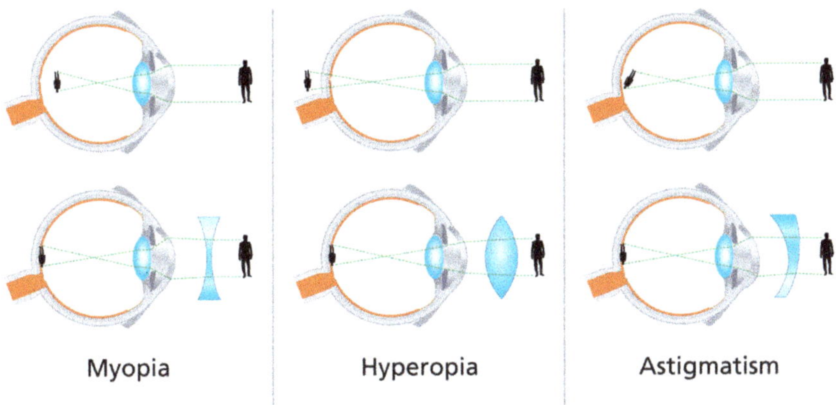

Figure 9. Illustration of different types of refractive error and their correction with lenses. Corneal refractive surgery changes the shape of the cornea according to the corrective lenses.

Refractive surgery consists of either reducing the refractive power of the cornea (by flattening) or by increasing its power (by steepening) or by modifying its curvature on a determined meridian to correct astigmatism (cylindrical correction).

3.1.1. Fs Flap Creation for Refractive Surgery

- LASIK

In the laser in situ keratomileusis (LASIK) procedure, a corneal flap is created. The flap is lifted and then excimer- or solid-state UV-laser energy is used to change the cornea's refractive power by flattening or steepening the stromal bed. Later, the flap is repositioned. Before the advent of fs-laser technology, the flap was created using mechanical devices called microkeratomes. With fs-laser technology, the flap can be completed in various patterns (Figure 10). Kezirian et al. compared fs-(IntraLase) created flaps to flaps with two different microkeratomes. They found in the fs group more predictable flap thickness, better astigmatic neutrality, and decreased epithelial injury [35]. Chen et al. confirmed the superiority of fs-laser-created flaps over those cut by microkeratomes. Therefore, in recent years, fs technology has superseded microkeratomes in preparing flaps for LASIK [36].

- Stromal keratophakia (additive refractive surgery)

Keratophakia as a means to sculpt corneal curvature by adding tissue has been studied since 1949 by Barraquer [37]. Because the quality of the cuts was inconsistent and reactive wound healing along the edges of the cut created additional scarring, it was largely abandoned. With advancements in femtosecond technology, new steps are being taken in the direction of keratophakia. For one, it is now possible to prepare an intrastromal pocket or stromal bed with greater precision. Secondly, new inlay materials are being developed. Current research is focusing on decellularizing and preserving extracted lenticules from lenticule extraction surgeries [38].

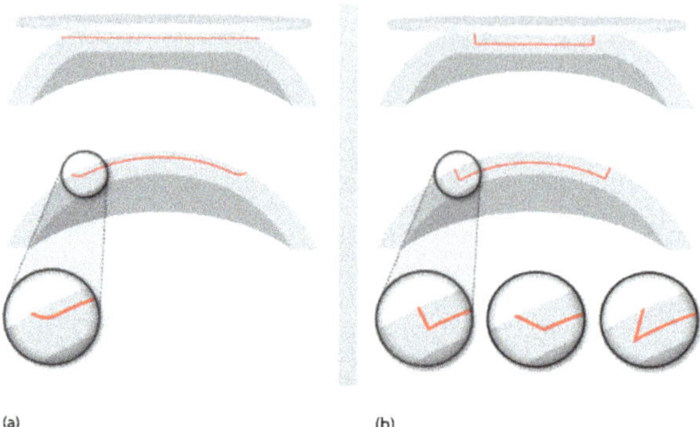

(a) (b)

Figure 10. Corneal flaps cut by fs laser: (**a**) straight plane (red) with continuously curved sides cut during vacuum docking to a flat interface, (**b**) angulated side cut options (3D cutting geometry).

3.1.2. Intrastromal Pockets

Multiple refractive surgery methods use fs-laser cuts to create "pocket"-shaped openings in the cornea, from which either material can be removed or implanted. In both cases, the refractive power of the cornea changes.

- Corneal stromal lenticule extraction

While many fs-associated surgical interventions in ophthalmology are merely improvements of pre-existing techniques, lenticule extraction is unique to fs-laser technology: the procedure was introduced in 2011 to treat myopia, and later also myopic astigmatism. It became known under the brand name "SMILE" (small incision lenticule extraction) of the Carl Zeiss Meditec AG. Later, other companies introduced their own laser systems for similar lenticule procedures under different brand names, including "SmartSight" by Schwind and "CLEAR" (corneal lenticule extraction for advanced refractive correction) by Ziemer Ophthalmic Systems AG.

The procedure is a "flapless" laser refractive technique that uses a single femtosecond laser system to create a pocket. The content of the pocket—the lenticule—is removed via a small access tunnel incision, and as a result, the cornea is flattened (see Figure 11). Instead of an almost 360-degree side cut, as in Lasik, lenticule extraction requires only a small arcuate cut of 50 degrees. Thereby more of the corneal nerves and Bowman layer remain untouched. In addition, sculpting the lenticule instead of ablating the same amount of tissue requires less laser energy. Therefore, the potential advantages of the lenticule technique over traditional laser in situ keratomileusis (LASIK) include reduced iatrogenic dry eye, a biomechanically stronger postoperative cornea with a smaller incision, and reduced laser energy required for refractive corrections [33,39–43]. However, the lenticule procedures have a steeper learning curve for surgeons, with potential complications related to lenticule dissection and removal, limitations with enhancements, and slower visual recovery in the initial phase (three months) [41]. Today, laser-refractive correction of hyperopia is not yet possible with lenticule extraction, but research in this field is ongoing. In a prospective, randomized paired-eye study, SMILE demonstrated good refractive outcomes in terms of predictability, efficacy, and safety. Since LASIK is reportedly an extremely safe and predictable procedure, it is unlikely to prove superiority with alternative methods, such as SMILE [44].

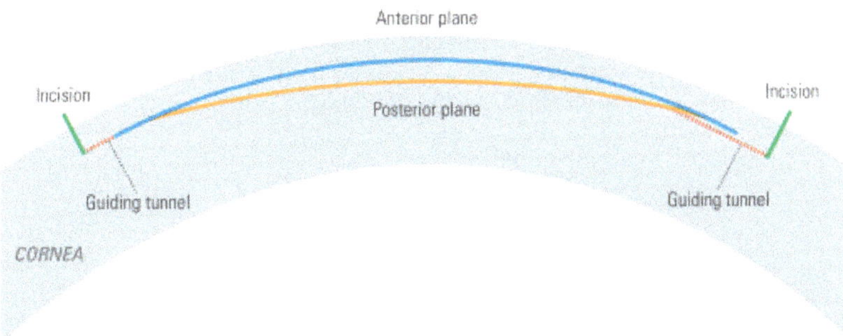

Figure 11. Schematic view of intrastromal lenticule cuts performed by an fs laser. The lenticule created between the anterior (blue) and posterior (yellow line) cut planes is extracted by the surgeon via an incision (green line). Optionally there is a second incision created to help mobilize the lenticule.

- Intrastromal corneal ring segments

Fs technology allows creating stromal pockets of specific size and shape at specific positions. Corneal ring segments (Figure 12) are placed into these pockets to change the curvature of the cornea, specifically in patients with thin and malleable corneas such as in keratoconus, a disease in which the central cornea becomes progressively deformed. Combining this procedure with a tissue strengthening intervention such as corneal crosslinking has been shown to improve uncorrected visual acuity in those keratoconus patients who do not tolerate contact lens correction [45].

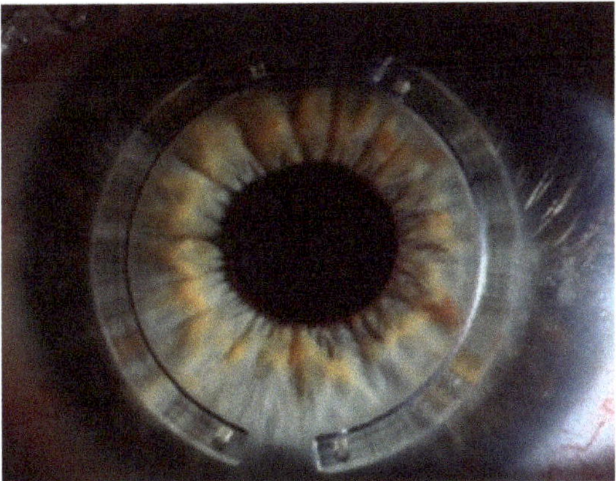

Figure 12. Intrastromal corneal ring segments implanted into fs-laser cut pockets.

3.1.3. Intrastromal and Trans-Stromal Cuts for Astigmatic Correction

The concept of corneal cuts for astigmatic correction was established more than 100 years ago [46] and underwent standardization in the late 1980s and 1990s [47,48]. Despite nomograms, there remained a significant level of unpredictability of the manually performed surgery. Astigmatic correction through toric intraocular lenses or corneal ablative surgery largely replaced correction through corneal stromal cuts. Fs-technology offers new opportunities to correct corneal astigmatism by means of corneal cuts: position, length, depth, curvature and the keratotomy angle can be put into practice with unprecedented precision and control. In addition, the fs-specific option of purely intrastromal keratotomies, decreases the potential risk of infection through gaping wounds. Although the general belief is that intrastromal cuts have less effect than transepithelial cuts, there are not enough data published for this relatively young technology to give evidence on significant differences in effectiveness between intrastromal and transepithelial cuts [49]. In general, fs-laser astigmatic correction is possible for both: smaller degrees of astigmatism in healthy corneas and larger astigmatic error in eyes with corneal pathology [50,51].

3.2. *Corneal Surgery*

3.2.1. Penetrating Keratoplasty

- Background

Keratoplasty (cornea transplantation) ranks among the oldest and most commonly performed human tissue transplantation types worldwide [52]. A corneal button from a deceased donor is sutured into the recipient cornea. Astigmatism is the leading cause of poor visual outcome after keratoplasty. The better the trephination (cut to separate the corneal button from the cornea) of donor and recipient, the better the fit between the transplant and the recipient and the least the astigmatism.

- Trephination

A perfect trephination system produces a congruent recipient bed and donor buttons and thereby allows well-centered tension-free fitting, and efficiently waterproof-adapting incision edges [53]. Different trephination systems are currently available: handheld, motor-trephine, excimer-laser, or fs-laser based. Comparison of motor-trephine and excimer-based trephination has shown better alignment of the graft in the recipient bed after excimer laser trephination [54].

It is often problematic to ensure proper centration with trephination in the recipient eye. Fs technology allows for perfect limbal oriented centration through OCT-visualization.

Another problem with trephination is the mechanism by which the recipient eye and donor button are fixated and stabilized; any mechanical impact on the tissue during trephination causes compression and distortion and will decrease the fit of recipient and donor (Figure 13). Common fixation mechanisms include vacuum and applanation and a combination of both (vacuum suction with applanation). While fs technology avoids some of the trephination pitfalls of mechanical trephination, comparison of fs- and excimer-assisted trephination showed nevertheless superiority of alignment in all sutures-out keratoplasty patients in the excimer group [55].

Figure 13. Comparison of donor and recipient trephination profiles: (**a**) applanation, (**b**) liquid optic interface.

Different stabilization systems could explain this superiority. While the excimer-assisted keratoplasty does not require applanation of the cornea, it is needed for the fs laser used in the cited studies. The new liquid optics interface assisted fs-Keratoplasty method could solve this problem: Here, cutting both recipient and donor is achieved within a liquid interface, which leaves the curvature of the cornea undisturbed. This reduces shear- and compression artifacts in the tissue and improves congruent fitting of the recipient and donor [56]. It will therefore be interesting to compare liquid optics interface fs-trephinations with excimer laser-assisted trephinations in the future.

- Sidecuts

In femtolaser-assisted keratoplasty (FLAK), different side-cut profiles can be chosen (Figure 14). Theoretical advantages include increased wound surface and thereby accelerated healing and wound stability, better vertical and horizontal alignment of the recipient and donor [57], preservation of healthy recipient corneal endothelium (mushroom), or transplantation of proportionally more endothelial cells with the top hat profile [58,59]. It remains to be seen if other factors, such as suture techniques, have to be modified to transmit these theoretical advantages into true clinical benefits [59].

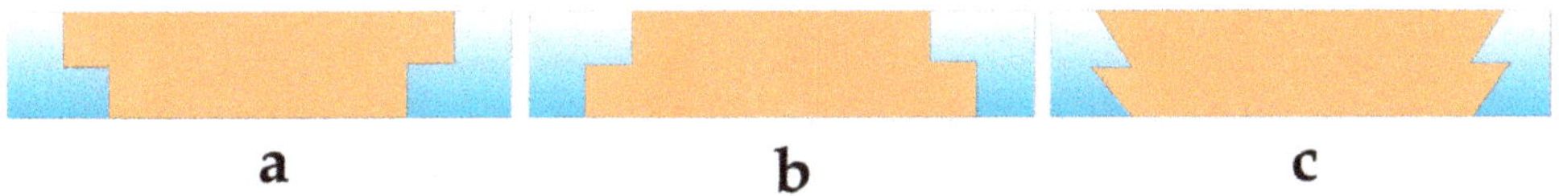

Figure 14. Sidecut profiles: (**a**) mushroom, (**b**) top hat, (**c**) zig-zag.

3.2.2. Lamellar Keratoplasty

- Background

The cornea is structured in five parallel layers. Often, not all layers of the cornea are diseased. Scars from trauma or infection commonly involve the anterior layers (Bowman layer, anterior stroma). In contrast, some inherited corneal dystrophies (i.e., inborn progressing tissue degeneration) affect only the inner most layers (Descemet's membrane and endothelium, see Figure 15). Selectively transplanting the pathological layers has several advantages: less tissue is transplanted, and thereby rejection is limited. With the scarcity of donor material, a donor button can theoretically be divided between two recipients. The integrity of the eye is less constrained. Since there is little adhesion between the interfaces of the corneal layers, manipulation at these levels is possible and visual results are excellent.

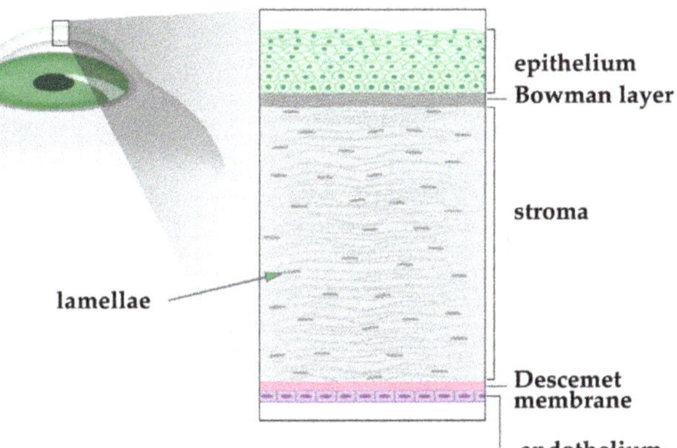

Figure 15. Illustration of corneal layers.

- Deep anterior lamellar keratoplasty (DALK)

In deep anterior lamellar keratoplasty (DALK), approximately 95% of the anterior corneal layers are removed, and only the innermost layers, Descemet's membrane, and endothelial cell layer stay behind [60]. It is possible to separate Descemet's membrane (DM) from the anterior stromal layers by air injection [61]. The surgical difficulty consists of finding the right entry-level for the air injection to initiate separation: too deep and DM is perforated, and the surgery has to be converted to a penetrating full thickness keratoplasty; too high and the air injection will not separate the layers, because only at the true interface is there minimal adhesion and the layers can be separated. Trials to create an fs-assisted cut in the pre-Descemet's stroma instead of separating the two layers led to lower visual clarity in comparison to true "layer separation" [62]. A new approach in fs technology resolves this dilemma—using OCT visualization, an fs-prepared channel is created that guides the cannula to the desired position and depth of the cornea (Figure 16). The surgeon can control the depth of the injection site individually adjusted to the thinnest point of the patient's cornea [63]. Buzzonetti et al. compared fs-DALK to mechanical DALK in 20 pediatric patients [64] and concluded that fs-assisted trephination could reduce the postoperative spherical equivalent amount. In conclusion, fs-assisted DALK can improve the success rate of big-bubble creation. By improving donor/recipient fit through fs-created sidecuts, the postoperative spherical equivalent is reduced, and healing accelerated.

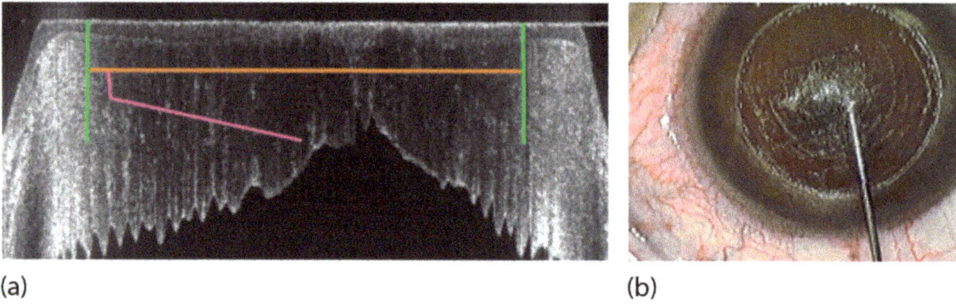

Figure 16. Deep anterior lamellar keratoplasty (DALK) procedure: (**a**) OCT-guided placement of the cuts: side cut in green, pre-Descemet's stroma cut in orange, guiding channel in pink. The posterior cornea is folded due to the applanating docking. (**b**) Insertion of the cannula for air injection through the precut guiding channel, just above the Descemet's membrane.

- Posterior lamellar keratoplasty

Posterior lamellar keratoplasty has revolutionized corneal transplant surgery in the past two decades: the cornea's clarity depends on healthy endothelial cells. These cells are thought of as non-regenerating highly specialized cells. In a disease called "Fuchs endothelial dystrophy", but also after traumatic or multiple surgical interventions, these cells cease to do their job in maintaining corneal clarity and patients eventually become blind. Transplanting these cells, be it with a small amount of corneal stroma, so-called Descemet stripping endothelial keratoplasty "DSEK", or with Descemet's membrane alone as the carrier, so-called Descemet's membrane endothelial keratoplasty "DMEK", reverses the process of corneal opacification and especially in the case of DMEK causes perfect visual acuity [60,65]. Part of the surgery consists of removing the old, nonfunctioning Descemet's membrane from the recipient cornea. This process is usually done by scraping and pulling the membrane manually. Sorkin et al. have published several papers on creating the descemetorhexis (cutting out of a part of the Descemet's membrane) with femtosecond laser assistance. The advantage is the perfect centration, shape, and size of the removed area [66–68]. It remains unclear if this advantage can solely be attributed to the fs-assisted descemetorhexis or if other causes, such as a better fit of transplanted and remaining Descemet's membrane, can explain these results. Nevertheless, it shows the immense versatility and breadth of applications that fs-laser technology provides in corneal surgery.

3.3. Cataract Surgery

Ophthalmic surgeons have been trying to implement laser in cataract surgery for decades. Bille and Schanzlin proposed ultrashort laser pulses for ablation of the cataractous lens in 1992 [69]. Nagy was the first reporting on the use of fs laser for cataract surgery [16]. There has been a quick evolution of the technology and platform ability since then by several manufacturers. Femtosecond laser technology reduces energy and precise laser application at a certain depth in tissue with minimal damage to adjacent areas. Currently, modern and commercially available fs systems allow the following steps to be taken over by the machine: (a) imaging and measurement of the anterior segment of the eye (incl. cornea, anterior chamber, iris, lens), (b) planning of fs-laser cut application to the tissue (including location depth, pattern, and size), (c) corneal incisions (full thickness for the introduction of instruments to the eye or partial thickness for treatment of corneal astigmatism), (d) circular incision to the anterior lens capsule (capsulotomy), and (e) fragmentation of the cataractous lens nucleus. For all of the abovementioned purposes, the eye must be fixed to laser optics by vacuum docking for precise laser application to the intended area and depth. While some systems use liquid optics interfaces (Ziemer Femto LDV), others have a curved applanation lens and suction system (LenSx, Alcon and Victus, Bausch and Lomb) or a fluid filled suction ring (Catalys, Johnson and Johnson and LensAR) [70]. The systems mentioned have all been CE marked and approved by the USA Food and Drug Administration for cataract surgery (Table 1).

Table 1. Overview of five commercially available femtosecond lasers for eye surgery.

	IntraLase (AMO, USA)	**Wavelight FS200 (Alcon, USA)**	**LenSx (Alcon, USA)**	**LensAR (LensAR, Topcon, USA)**	**Catalys (Johnson and Johnson, USA)**	**Victus (Bausch and Lomb, Germany)**	**VisuMax (Zeiss Meditec, Germany)**	**LDV Z8 (Ziemer, Switzerland)**	**Atos * (Schwind, Germany)**
Pulse repetition rate (kHz)	30–150	200	60	80	120	80/160	500	10,000	<to 4000
Pulse duration (fs)	>500	350	600–800	500	<600	290–550	220–580	250	<295
Pulse energy (μJ)	Ca. 1	<1.5	>15	7–15	3–10	6–10	<1	<1	<1
Applications:									
LASIK flaps	x	x		x		x	x	x	x
Refractive Lenticules							x	x	x
Cornea Surgery	x	x				x		x	
Cataract Surgery			x	x	x	x		x	
Patient interface	Flat applan. interface	Flat applan. interface	Curved softfit interface	Fluid filled interface	Liquid interface	Semiliquid curved interface	Curved interface	Liquid and flat interfaces	Curved interface

*: according to manufacturer.

The main advantages of fs-assisted cataract surgeries are the precision and repeatability of laser incisions to the tissue; reduction in ultrasound energy used for emulsification (liquification) of the lens nucleus by precutting it into small pieces; perfect sizing of corneal incision with regards to position, length and depth; and predictability in capsulotomy size and position. Despite the aforementioned obvious advantages and numerous studies showing superiority in performing the single surgical steps over the ones manually performed by surgeons, meta-analysis studies could not prove overall outcome advantages of fs-laser-assisted surgery versus the conventional phacoemulsification manual operation [71,72]. A randomized multicenter clinical trial including 1476 eyes of 907 patients could not prove superiority of fs-assisted cataract surgery over the traditional phacoemulsification method [73]. Nevertheless, the authors report no severe adverse events during the fs-laser procedure. Similarly, review articles emphasize usefulness of fs-assisted cataract surgery in some patient groups, i.e., those with low corneal endothelial cell counts, but a clear advantage of the fs method over manual phacoemulsification is not reported in routine cases [74,75]. Furthermore, the authors question the cost-effectiveness of fs-assisted surgery.

The question remains: Why should a high-precision system not be superior to the manual? Two answers merit consideration. (1) Several studies were done using the first-generation femtosecond laser systems comparing those to the conventional phacoemulsification surgery that has undergone evolution and perfection for several decades. Results of comparative studies using the newest laser devices could give a better comparison. (2) In most comparative studies, the conventional surgery has been performed by outstanding, high-volume, and exceptionally talented cataract surgeons. Comparing the visual outcomes of such surgeons to the machine results is like assessing the abilities and outcomes of first-generation autonomous driving systems to those of Formula-1 drivers. In the following, we describe each step of the cataract surgery taken over by the fs machine.

3.3.1. Capsulotomy

Traditionally, cataract surgeons access the cataractous lens by manually opening the anterior lens capsule by pulling in a continuous curvilinear manner. This maneuver is performed using a needle or forceps and is called capsulorhexis. While the hard inner part of the lens is removed, the outer capsular bag is maintained. This bag is used as a mounting plate to fixate an implanted acrylic intraocular lens at the same position from where the cataractous lens was removed. Size, position, and shape of the capsulorhexis are related to the effective lens position, a determinant of the intraocular lens (IOL) power. The IOL power determines the postoperative refractive error of the eye. Inappropriate sizing of the capsular opening may result in IOL tilt, decentration, and increased posterior lens capsule opacification [76–78]. Perfect lens position is of particular importance to the IOLs with complex optical properties, e.g., multifocal, toric (for astigmatism correction), or those with extended depth of focus [79,80].

While a perfect capsular opening is of the highest significance for surgical success, it is one of the most challenging maneuvers in cataract surgery. The learning curve of surgeons in training can be quite flat. At the same time, even in the hands of the most experienced surgeons, the capsulorhexis can be unpredictable and perfect sizing at the submillimeter precision appears only possible with machines performing the step. Furthermore, the manual outcome is dependent on the axial length of the eye, pupil size, image enlargement by the cornea [81], and inherent features of the individual eye, for example, true exfoliation of the lens capsule (an eye disease) [82]. Femtosecond lasers overcome all these challenges by creating precise, predictable, repeatable, well-centered capsular opening, called laser capsulotomy, even in challenging cases with loose zonules (tissue fibers holding the lens in place), pediatric or mature cataracts, shallow anterior chamber, etc. Machine superiority has been demonstrated in several studies [78,83]. Another potential advantage of laser capsulotomy is centration of the opening on the eye's true optical axis or the lens apex instead of centration on the pupil, as is usually done in manual capsulorhexis. Furthermore, innovative IOLs are available that are dependent on perfect capsulotomy sizing at a submillimeter level. Those designs allow IOL centration based on the capsulotomy rather than on the capsular bag [84].

3.3.2. Nucleus Fragmentation

The human lens loses transparency and flexibility throughout life. A cataractous lens cannot be removed through a small incision by suction alone; rather, emulsification or fragmentation of the hard lens nucleus is necessary. Conventional cataract surgery uses ultrasound energy within the eye to liquefy the lens nucleus. Femtosecond laser technology allows precutting the nucleus in almost any imaginable shape and reduces the ultrasound energy needed for emulsification. This is an advantage as the ultrasound energy is a cause of oxidative stress, heat, and inflammation, and damage to the tissue [85]. The most susceptible tissue is the one-layer cell sheet of the corneal endothelium, which is of utmost importance to corneal transparency. Studies have shown less endothelial cell loss and decreased corneal swelling when using fs technology [86–88]. The importance of protection to the corneal endothelium becomes clear when we consider that corneal endothelial cells do not multiply after injury. The only way to repair the damage once clinically relevant is lamellar corneal grafting (transplantation). Several lens fragmentation patterns have been introduced, and currently, it is not yet clear which design to prefer in a particular clinical setting. Future studies should focus on the optimal fs lens nucleus fragmentation pattern to reduce effective ultrasound energy used intraoperatively.

3.3.3. Corneal Incisions

- Full-thickness incisions

Full-thickness incisions through the cornea are necessary for the introduction of instruments into the eye. Traditionally, a metal scalpel or diamonds are used for creating them in different sizes. Fs technology allows predictable sizing (width, length, and depth) of full-thickness corneal incisions. Since these incisions need to be self-sealing, a perfect wound architecture incision is mandatory to prevent wound leakage postoperatively. Incorrect positioning of the wound induces astigmatism and can provoke prolapse of the iris during the surgery. Studies have shown increased repeatability and safety of wound construction using fs technology, resulting in higher stability and water tightness [89–91].

- Partial-thickness incisions

Partial-thickness incisions into the cornea help to reduce preoperatively existing corneal astigmatism. Several nomograms have been developed, addressing length, position, and depth of the incisions for different amounts of astigmatism. Fs technology allows higher predictability and repeatability of partial thickness incisions, or even completely intrastromal corneal incisions [50]. Since the incision depths are up to 90% of the corneal thickness, laser precision helps prevent inadvertent penetration, as reported in manually performed antiastigmatic keratotomies. Fs-laser-assisted corneal incision could be as safe and effective as toric IOLs to reduce astigmatism [92]. Future studies will show whether predictability, safety, and efficacy of fs-laser-assisted keratotomies can be further improved by implementing nonperpendicular incision directions.

3.3.4. Future Applications

Probably the most important evolutionary step in ophthalmic fs-laser devices will be miniaturization, mobility, and versatility. Tools available soon are fs laser-assisted primary posterior capsulotomy and lens capsule marking for positioning of toric IOL. On the horizon, another technology involves changing the IOL power postoperatively through fs-laser energy to achieve emmetropia in all eyes [93].

4. Summary

In summary, fs-laser technology has evolved over the past decades into a precise and reliable tool in ophthalmic surgery. While some of the applications have not finished evolving and require further research and development, fs-laser-assisted cataract and corneal surgery have reached highly standardized levels worldwide. For these surgeries, fs-laser technology has improved patient safety and clinical outcomes and opened gateways to new surgical approaches.

Author Contributions: Conceptualization, C.L., T.A. and A.M.; validation, C.R.; writing—original draft preparation, C.L., T.A. and A.M.; writing—review and editing, T.A., A.M., C.R. and C.L.; supervision, C.R. All authors have read and agreed to the published version of the manuscript.

Funding: This research received no external funding.

Conflicts of Interest: A.M. is a consultant to Ziemer Ophthalmics, CH-2562 Port, Switzerland.

References

1. Welch, A.J.; van Gemert, M. *Optical-Thermal Response of Laser-Irradiated Tissue*; Springer: Berlin/Heidelberg, Germany, 2011.
2. Kaschke, M.; Donnerhacke, K.-H.; Rill, M.S. *Optical Devices in Ophthalmology and Optometry: Technology, Design Principles and Clinical Applications*; WILEY-VCH Verlag GmbH & Co. KGaA: Weinheim, Germany, 2014.
3. Aron-Rosa, D.; Aron, J.J.; Griesemann, M.; Thyzel, R. Use of the neodymium-yag laser to open the posterior capsule after lens implant surgery: A preliminary report. *Am. Intra-Ocular Implant. Soc. J.* **1980**, *6*, 352–354. [CrossRef]
4. Stern, D.; Schoenlein, R.W.; Puliafito, C.A.; Dobi, E.T.; Birngruber, R.; Fujimoto, J.G. Corneal Ablation by Nanosecond, Picosecond, and Femtosecond Lasers at 532 and 625 nm. *Arch. Ophthalmol.* **1989**, *107*, 587–592. [CrossRef] [PubMed]
5. Huang, D.; Swanson, E.A.; Lin, C.P.; Schuman, J.S.; Stinson, W.G.; Chang, W.; Hee, M.R.; Flotte, T.; Gregory, K.; Puliafito, C.A.; et al. Optical Coherence Tomography. *Science* **1991**, *254*, 1178–1181. [CrossRef]
6. Bhargava, R.; Kumar, P.; Phogat, H.; Chaudhary, K.P. Neodymium-yttrium aluminium garnet laser capsulotomy energy levels for posterior capsule opacification. *J. Ophthalmic Vis. Res.* **2015**, *10*, 37–42. [CrossRef] [PubMed]
7. Vogel, A.; Noack, J.; Hüttman, G.; Paltauf, G. Mechanisms of femtosecond laser nanosurgery of cells and tissues. *Appl. Phys. A* **2005**, *81*, 1015–1047. [CrossRef]
8. Heisterkamp, A.; Ripken, T.; Lubatschowski, H.; Welling, H.; Luetkefels, E.; Drommer, W.; Ertmer, W. Intrastromal cutting effects in rabbit cornea using femtosecond laser pulses. In *Optical Biopsy and Tissue Optics*; International Society for Optics and Photonics: Bellingham, WA, USA, 2000; pp. 52–60.
9. Pepose, J.S.; Lubatschowski, H. Comparing Femtosecond Lasers. *Cataract Refract. Surg. Today* **2008**, *10*, 45–51.
10. Vogel, A.; Busch, R. Shock Wave Emission and Cavitation Bubble Generation by Picosecond and Nanosecond Optical Breakdown in Water. *J. Acoust. Soc. Am.* **1996**, *100*, 148–165. [CrossRef]
11. Vogel, A.; Venugopalan, V. Mechanisms of Pulsed Laser Ablation of Biological Tissues. *Chem. Rev.* **2003**, *103*, 577–644. [CrossRef]
12. Tinne, N.; Knoop, G.; Kallweit, N.; Veith, S.; Bleeker, S.; Lubatschowski, H.; Krüger, A.; Ripken, T. Effects of cavitation bubble interaction with temporally separated fs-laser pulses. *J. Biomed. Opt.* **2014**, *19*, 48001. [CrossRef]
13. Lubatschowski, H.; Maatz, G.; Heisterkamp, A.; Hetzel, U.; Drommer, W.; Welling, H.; Ertmer, W. Application of ultrashort laser pulses for intrastromal refractive surgery. *Graefe's Arch. Clin. Exp. Ophthalmol.* **2000**, *238*, 33–39. [CrossRef]
14. Ratkay-Traub, I.; Juhasz, T.; Horvath, C.; Suarez, C.; Kiss, K.; Ferincz, I.; Kurtz, R. Ultra-short pulse (femtosecond) laser surgery: Initial use in LASIK flap creation. *Ophthalmol. Clin. N. Am.* **2001**, *14*, 347–355.
15. Nagy, Z.Z.; McAlinden, C. Femtosecond laser cataract surgery. *Eye Vis.* **2015**, *2*, 1–8. [CrossRef] [PubMed]
16. Nagy, Z.; Takacs, A.; Filkorn, T.; Sarayba, M. Initial Clinical Evaluation of an Intraocular Femtosecond Laser in Cataract Surgery. *J. Refract. Surg.* **2009**, *25*, 1053–1060. [CrossRef] [PubMed]
17. Ostovic, M.; Klaproth, O.K.; Hengerer, F.H.; Mayer, W.J.; Kohnen, T. Light Microscopy and Scanning Electron Microscopy Analysis of Rigid Curved Interface Femtosecond Laser-Assisted and Manual Anterior Capsulotomy. *J. Cataract Refract. Surg.* **2013**, *39*, 1587–1592. [CrossRef]
18. Lubatschowski, H. Overview of Commercially Available Femtosecond Lasers in Refractive Surgery. *J. Refract. Surg.* **2008**, *24*, S102–S107. [CrossRef]
19. Riau, A.K.; Liu, Y.C.; Lwin, N.C.; Ang, H.P.; Tan, N.Y.; Yam, G.H.; Tan, D.T.; Mehta, J.S. Comparative Study of Nj- and Muj-Energy Level Femtosecond Lasers: Evaluation of Flap Adhesion Strength, Stromal Bed Quality, and Tissue Responses. *Invest. Ophthalmol. Vis. Sci.* **2014**, *55*, 3186–3194. [CrossRef]
20. Mayer, W.J.; Klaproth, O.K.; Ostovic, M.; Terfort, A.; Vavaleskou, T.; Hengerer, F.H.; Kohnen, T. Cell Death and Ultrastructural Morphology of Femtosecond Laser–Assisted Anterior Capsulotomy. *Investig. Opthalmol. Vis. Sci.* **2014**, *55*, 893–898. [CrossRef]
21. Schultz, T.; Joachim, S.C.; Stellbogen, M.; Dick, H.B. Prostaglandin Release During Femtosecond Laser-Assisted Cataract Surgery: Main Inducer. *J. Refract. Surg.* **2015**, *31*, 78–81. [CrossRef]
22. Schwarzenbacher, L.; Schartmueller, D.; Leydolt, C.; Menapace, R. Intra-individual comparison of cytokine and prostaglandin levels with and without low-energy, high-frequency femtosecond laser cataract pretreatment following single-dose topical NSAID application. *J. Cataract Refract. Surg.* **2020**, *46*, 1086–1091. [CrossRef]
23. Fercher, A.F.; Drexler, W.; Hitzenberger, C.K.; Lasser, T. Optical Coherence Tomography—Principles and Applications. *Rep. Prog. Phys.* **2003**, *66*, 239–303. [CrossRef]
24. Schuman, J.S.; Puliafito, C.A.; Fujimoto, J.G.; Duker, J.S. *Optical Coherence Tomography of Ocular Diseases*; Slack Inc.: San Francisco, CA, USA, 2012.
25. Wojtkowski, M.; Leitgeb, R.A.; Kowalczyk, A.; Bajraszewski, T.; Fercher, A.F. In vivo human retinal imaging by Fourier domain optical coherence tomography. *J. Biomed. Opt.* **2002**, *7*, 457–463. [CrossRef] [PubMed]
26. Izatt, J.A.; Hee, M.R.; Swanson, E.A.; Lin, C.P.; Huang, D.; Schuman, J.S.; Puliafito, C.A.; Fujimoto, J.G. Micrometer-Scale Resolution Imaging of the Anterior Eye in Vivo with Optical Coherence Tomography. *Arch. Ophthalmol.* **1994**, *112*, 1584–1589. [CrossRef] [PubMed]
27. Kermani, O.; Fabian, W.; Lubatschowski, H. Real-Time Optical Coherence Tomography-Guided Femtosecond Laser Sub-Bowman Keratomileusis on Human Donor Eyes. *Am. J. Ophthalmol.* **2008**, *146*, 42–45. [CrossRef] [PubMed]

28. Chang, J.S.; Chen, I.N.; Chan, W.M.; Ng, J.C.; Chan, V.K.; Law, A.K. Initial evaluation of a femtosecond laser system in cataract surgery. *J. Cataract Refract. Surg.* **2014**, *40*, 29–36. [CrossRef] [PubMed]
29. Grewal, D.S.; Schultz, T.; Basti, S.; Dick, H.B. Femtosecond laser–assisted cataract surgery—Current status and future directions. *Surv. Ophthalmol.* **2016**, *61*, 103–131. [CrossRef]
30. Talamo, J.H.; Gooding, P.; Angeley, D.; Culbertson, W.W.; Schuele, G.; Andersen, D.; Marcellino, G.; Essock-Burns, E.; Batlle, J.; Feliz, R.; et al. Optical patient interface in femtosecond laser–assisted cataract surgery: Contact corneal applanation versus liquid immersion. *J. Cataract Refract. Surg.* **2013**, *39*, 501–510. [CrossRef]
31. Mirshahi, A.; Kohnen, T. Effect of Microkeratome Suction During LASIK on Ocular Structures. *Ophthalmology* **2005**, *112*, 645–649. [CrossRef]
32. Flaxel, C.J.; Choi, Y.H.; Sheety, M.; Oeinck, S.C.; Lee, J.Y.; McDonnell, P.J. Proposed Mechanism for Retinal Tears after Lasik: An Experimental Model. *Ophthalmology* **2004**, *111*, 24–27. [CrossRef]
33. Kanclerz, P.; Grzybowski, A. Does Corneal Refractive Surgery Increase the Risk of Retinal Detachment? A Literature Review and Statistical Analysis. *J. Refract. Surg.* **2019**, *35*, 517–524. [CrossRef]
34. Toth, C.A.; Mostafavi, R.; Fekrat, S.; Kim, T. LASIK and vitreous pathology after LASIK. *Ophthalmology* **2002**, *109*, 624–625. [CrossRef]
35. Kezirian, G.M.; Stonecipher, K.G. Comparison of the IntraLase femtosecond laser and mechanical keratomes for laser in situ keratomileusis. *J. Cataract Refract. Surg.* **2004**, *30*, 804–811. [CrossRef] [PubMed]
36. Chen, S.; Feng, Y.; Stojanovic, A.; Jankov, M.R., II; Wang, Q. Intralase Femtosecond Laser Vs Mechanical Microkeratomes in Lasik for Myopia: A Systematic Review and Meta-Analysis. *J. Refract. Surg.* **2012**, *28*, 15–24. [CrossRef] [PubMed]
37. Barraquer, J.I. The History and Evolution of Keratomileusis. *Int. Ophthalmol. Clin.* **1996**, *36*, 1–7. [CrossRef] [PubMed]
38. Riau, A.K.; Liu, Y.-C.; Yam, G.H.; Mehta, J.S. Stromal keratophakia: Corneal inlay implantation. *Prog. Retin. Eye Res.* **2020**, *75*, 100780. [CrossRef]
39. Wei, S.; Wang, Y. Comparison of corneal sensitivity between FS-LASIK and femtosecond lenticule extraction (ReLEx flex) or small-incision lenticule extraction (ReLEx smile) for myopic eyes. *Graefe's Arch. Clin. Exp. Ophthalmol.* **2013**, *251*, 1645–1654. [CrossRef]
40. Ang, M.; Farook, M.; Htoon, H.M.; Mehta, J.S. Randomized Clinical Trial Comparing Femtosecond LASIK and Small-Incision Lenticule Extraction. *Ophthalmology* **2020**, *127*, 724–730. [CrossRef]
41. Ang, M.; Mehta, J.S.; Chan, C.; Htoon, H.M.; Koh, J.C.; Tan, D. Refractive lenticule extraction: Transition and comparison of 3 surgical techniques. *J. Cataract Refract. Surg.* **2014**, *40*, 1415–1424. [CrossRef]
42. Riau, A.K.; Angunawela, R.I.; Chaurasia, S.S.; Lee, W.S.; Tan, D.T.; Mehta, J.S. Early Corneal Wound Healing and Inflammatory Responses after Refractive Lenticule Extraction (ReLEx). *Investig. Opthalmol. Vis. Sci.* **2011**, *52*, 6213–6221. [CrossRef]
43. Vestergaard, A.H.; Grauslund, J.; Ivarsen, A.; Hjortdal, J. Efficacy, safety, predictability, contrast sensitivity, and aberrations after femtosecond laser lenticule extraction. *J. Cataract Refract. Surg.* **2014**, *40*, 403–411. [CrossRef]
44. Sandoval, H.P.; Donnenfeld, E.D.; Kohnen, T.; Lindstrom, R.L.; Potvin, R.; Tremblay, D.M.; Solomon, K.D. Modern laser in situ keratomileusis outcomes. *J. Cataract Refract. Surg.* **2016**, *42*, 1224–1234. [CrossRef]
45. Hashemi, H.; Alvani, A.; Seyedian, M.A.; Yaseri, M.; Khabazkhoob, M.; Esfandiari, H. Appropriate Sequence of Combined Intracorneal Ring Implantation and Corneal Collagen Cross-Linking in Keratoconus: A Systematic Review and Meta-Analysis. *Cornea* **2018**, *37*, 1601–1607. [CrossRef] [PubMed]
46. Lans, L.J. Experimentelle Untersuchungen über Entstehung von Astigmatismus Durch Nicht-Perforirende Corneawunden. *Albrecht Graefes Arch. Ophthalmol.* **1898**, *45*, 117–152. [CrossRef]
47. Osher, R.H. Paired transverse relaxing keratotomy: A combined technique for reducing astigmatism. *J. Cataract Refract. Surg.* **1989**, *15*, 32–37. [CrossRef]
48. Thornton, S.P.; Sanders, D.R. Graded Nonintersecting Transverse Incisions for Correction of Idiopathic Astigmatism. *J. Cataract Refract. Surg.* **1987**, *13*, 27–31. [CrossRef]
49. Chang, J.S.M. Femtosecond laser-assisted astigmatic keratotomy: A review. *Eye Vis.* **2018**, *5*, 6. [CrossRef] [PubMed]
50. Mirshahi, A.; Latz, C. Femtosecond laser-assisted astigmatic keratotomy. *Ophthalmologe* **2020**, *117*, 415–423. [CrossRef]
51. Chan, T.C.-Y.; Ng, A.L.; Cheng, G.P.; Wang, Z.; Woo, V.C.; Jhanji, V. Corneal Astigmatism and Aberrations After Combined Femtosecond-Assisted Phacoemulsification and Arcuate Keratotomy: Two-Year Results. *Am. J. Ophthalmol.* **2016**, *170*, 83–90. [CrossRef]
52. Tan, D.; Dart, J.K.G.; Holland, E.J.; Kinoshita, S. Corneal transplantation. *Lancet* **2012**, *379*, 1749–1761. [CrossRef]
53. Seitz, B.; Langenbucher, A.; Hager, T.; Janunts, E.; El-Husseiny, M.; Szentmary, N. Penetrating Keratoplasty for Keratoconus—Excimer Versus Femtosecond Laser Trephination. *Open Ophthalmol. J.* **2017**, *11*, 225–240. [CrossRef]
54. Seitz, B.; Hager, T.; Langenbucher, A.; Naumann, G.O.H. Reconsidering Sequential Double Running Suture Removal after Penetrating Keratoplasty: A Prospective Randomized Study Comparing Excimer Laser and Motor Trephination. *Cornea* **2018**, *37*, 301–306. [CrossRef]
55. Tóth, G.; Szentmáry, N.; Langenbucher, A.; Akhmedova, E.; El-Husseiny, M.; Seitz, B. Comparison of Excimer Laser Versus Femtosecond Laser Assisted Trephination in Penetrating Keratoplasty: A Retrospective Study. *Adv. Ther.* **2019**, *36*, 3471–3482. [CrossRef] [PubMed]

56. Boden, K.T.; Schlosser, R.; Boden, K.; Januschowski, K.; Szurman, P.; Rickmann, A. Novel Liquid Interface for Femtosecond Laser-Assisted Penetrating Keratoplasty. *Curr. Eye Res.* **2020**, *45*, 1051–1057. [CrossRef] [PubMed]
57. Maier, P.; Böhringer, D.; Birnbaum, F.; Reinhard, T. Improved Wound Stability of Top-Hat Profiled Femtosecond Laser-Assisted Penetrating Keratoplasty In Vitro. *Cornea* **2012**, *31*, 963–966. [CrossRef] [PubMed]
58. Seitz, B.; Brünner, H.; Viestenz, A.; Hofmann-Rummelt, C.; Schlötzer-Schrehardt, U.; Naumann, G.O.; Langenbucher, A. Inverse Mushroom-Shaped Nonmechanical Penetrating Keratoplasty Using a Femtosecond Laser. *Am. J. Ophthalmol.* **2005**, *139*, 941–944. [CrossRef] [PubMed]
59. Maier, P.C.; Birnbaum, F.; Reinhard, T. Therapeutic Applications of the Femtosecond Laser in Corneal Surgery. *Klin. Monbl. Augenheilkd.* **2010**, *227*, 453–459. [CrossRef] [PubMed]
60. Reinhart, W.J.; Musch, D.C.; Jacobs, D.S.; Lee, W.B.; Kaufman, S.C.; Shtein, R.M. Deep Anterior Lamellar Keratoplasty as an Alternative to Penetrating Keratoplasty a Report by the American Academy of Ophthalmology. *Ophthalmology* **2011**, *118*, 209–218. [CrossRef] [PubMed]
61. Anwar, M.; Teichmann, K.D. Big-Bubble Technique to Bare Descemet's Membrane in Anterior Lamellar Keratoplasty. *J. Cataract Refract. Surg.* **2002**, *28*, 398–403. [CrossRef]
62. De Macedo, J.P.; de Oliveira, L.A.; Hirai, F.; de Sousa, L.B. Femtosecond Laser-Assisted Deep Anterior Lamellar Keratoplasty in Phototherapeutic Keratectomy Versus the Big-Bubble Technique in Keratoconus. *Int. J. Ophthalmol.* **2018**, *11*, 807–812.
63. Buzzonetti, L.; Laborante, A.; Petrocelli, G. Standardized Big-Bubble Technique in Deep Anterior Lamellar Keratoplasty Assisted by the Femtosecond Laser. *J. Cataract Refract. Surg.* **2010**, *36*, 1631–1636. [CrossRef]
64. Buzzonetti, L.; Petrocelli, G.; Valente, P.; Petroni, S.; Parrilla, R.; Iarossi, G. Refractive Outcome of Keratoconus Treated by Big-Bubble Deep Anterior Lamellar Keratoplasty in Pediatric Patients: Two-Year Follow-up Comparison between Mechanical Trephine and Femtosecond Laser Assisted Techniques. *Eye Vis.* **2019**, *6*, 1. [CrossRef]
65. Stuart, A.J.; Romano, V.; Virgili, G.; Shortt, A.J. Descemet's Membrane Endothelial Keratoplasty (Dmek) Versus Descemet's Stripping Automated Endothelial Keratoplasty (Dsaek) for Corneal Endothelial Failure. *Cochrane Database Syst. Rev.* **2018**, *6*, d012097. [CrossRef] [PubMed]
66. Einan-Lifshitz, A.; Sorkin, N.; Boutin, T.; Showail, M.; Borovik, A.; Alobthani, M.; Chan, C.C.; Rootman, D.S. Comparison of Femtosecond Laser-Enabled Descemetorhexis and Manual Descemetorhexis in Descemet Membrane Endothelial Keratoplasty. *Cornea* **2017**, *36*, 767–770. [CrossRef] [PubMed]
67. Sorkin, N.; Mednick, Z.; Einan-Lifshitz, A.; Trinh, T.; Santaella, G.; Telli, A.; Chan, C.C.; Rootman, D.S. Three-Year Outcome Comparison Between Femtosecond Laser-Assisted and Manual Descemet Membrane Endothelial Keratoplasty. *Cornea* **2019**, *38*, 812–816. [CrossRef]
68. Sorkin, N.; Mimouni, M.; Santaella, G.; Trinh, T.; Cohen, E.; Einan-Lifshitz, A.; Chan, C.C.; Rootman, D.S. Comparison of Manual and Femtosecond Laser-Assisted Descemet Membrane Endothelial Keratoplasty for Failed Penetrating Keratoplasty. *Am. J. Ophthalmol.* **2020**, *214*, 1–8. [CrossRef] [PubMed]
69. Bille, J.F.; Schanzlin, D. Method for Removing Cataractous Material. US Patent US5246435A, 1993.
70. Boden, K.T.; Szurman, P. Current Value of Femtosecond Laser-Assisted Cataract Surgery. *Ophthalmologe* **2020**, *117*, 405–414. [CrossRef]
71. Wang, J.; Su, F.; Wang, Y.; Chen, Y.; Chen, Q.; Li, F. Intra and Post-Operative Complications Observed with Femtosecond Laser-Assisted Cataract Surgery Versus Conventional Phacoemulsification Surgery: A Systematic Review and Meta-Analysis. *BMC Ophthalmol.* **2019**, *19*, 177. [CrossRef]
72. Day, A.C.; Gore, D.M.; Bunce, C.; Evans, J.R. Laser-assisted cataract surgery versus standard ultrasound phacoemulsification cataract surgery. *Cochrane Database Syst. Rev.* **2016**, *7*, CD010735. [CrossRef]
73. Schweitzer, C.; Brezin, A.; Cochener, B.; Monnet, D.; Germain, C.; Roseng, S.; Sitta, R.; Maillard, A. Femtosecond Laser-Assisted Versus Phacoemulsification Cataract Surgery (Femcat): A Multicentre Participant-Masked Randomised Superiority and Cost-Effectiveness Trial. *Lancet* **2020**, *395*, 212–224. [CrossRef]
74. Kanclerz, P.; Alio, J.L. The benefits and drawbacks of femtosecond laser-assisted cataract surgery. *Eur. J. Ophthalmol.* **2020**, 1–10. [CrossRef]
75. Hooshmand, J.; Vote, B.J. Femtosecond Laser-Assisted Cataract Surgery, Technology, Outcome, Future Directions and Modern Applications. *Asia Pac. J. Ophthalmol.* **2017**, *6*, 393–400.
76. Sanders, D.R.; Higginbotham, R.W.; Opatowsky, I.E.; Confino, J. Hyperopic shift in refraction associated with implantation of the single-piece Collamer intraocular lens. *J. Cataract Refract. Surg.* **2006**, *32*, 2110–2112. [CrossRef] [PubMed]
77. Baumeister, M.; Bühren, J.; Kohnen, T. Tilt and decentration of spherical and aspheric intraocular lenses: Effect on higher-order aberrations. *J. Cataract Refract. Surg.* **2009**, *35*, 1006–1012. [CrossRef] [PubMed]
78. Friedman, N.J.; Palanker, D.; Schuele, G.; Andersen, D.; Marcellino, G.; Seibel, B.S.; Batlle, J.; Feliz, R.; Talamo, J.H.; Blumenkranz, M.S.; et al. Femtosecond laser capsulotomy. *J. Cataract Refract. Surg.* **2011**, *37*, 1189–1198. [CrossRef] [PubMed]
79. Kránitz, K.; Miháltz, K.; Sándor, G.L.; Takacs, A.; Knorz, M.C.; Nagy, Z.Z. Intraocular Lens Tilt and Decentration Measured By Scheimpflug Camera Following Manual or Femtosecond Laser–created Continuous Circular Capsulotomy. *J. Refract. Surg.* **2012**, *28*, 259–263. [CrossRef] [PubMed]

80. Kránitz, K.; Takacs, A.; Mih____ltz, K.; Kovács, I.; Knorz, M.C.; Nagy, Z.Z. Femtosecond Laser Capsulotomy and Manual Continuous Curvilinear Capsulorrhexis Parameters and Their Effects on Intraocular Lens Centration. *J. Refract. Surg.* **2011**, *27*, 558–563. [CrossRef] [PubMed]
81. Nagy, Z.Z.; Kránitz, K.; Takacs, A.I.; Mihaltz, K.; Kovács, I.; Knorz, M.C. Comparison of Intraocular Lens Decentration Parameters After Femtosecond and Manual Capsulotomies. *J. Refract. Surg.* **2011**, *27*, 564–569. [CrossRef]
82. Latz, C.; Mirshahi, A. Double Ring Sign of the Lens Capsule: Intraoperative Observation During Cataract Surgery. *Ophthalmologe* **2020**. [CrossRef]
83. Abouzeid, H.; Ferrini, W. Femtosecond-laser assisted cataract surgery: A review. *Acta Ophthalmol.* **2014**, *92*, 597–603. [CrossRef]
84. Holland, D.; Rufer, F. New Intraocular Lens Designs for Femtosecond Laser-Assisted Cataract Operations: Chances and Benefits. *Ophthalmologe* **2020**, *117*, 424–430. [CrossRef]
85. Murano, N.; Ishizaki, M.; Sato, S.; Fukuda, Y.; Takahashi, H. Corneal Endothelial Cell Damage by Free Radicals Associated with Ultrasound Oscillation. *Arch. Ophthalmol.* **2008**, *126*, 816–821. [CrossRef]
86. Abell, R.G.; Kerr, N.M.; Vote, B.J. Toward Zero Effective Phacoemulsification Time Using Femtosecond Laser Pretreatment. *Ophthalmology* **2013**, *120*, 942–948. [CrossRef] [PubMed]
87. Conrad-Hengerer, I.; Al Juburi, M.; Schultz, T.; Hengerer, F.H.; Dick, B.H. Corneal endothelial cell loss and corneal thickness in conventional compared with femtosecond laser–assisted cataract surgery: Three-month follow-up. *J. Cataract Refract. Surg.* **2013**, *39*, 1307–1313. [CrossRef] [PubMed]
88. Takacs, A.I.; Kovacs, I.; Mihaltz, K.; Filkorn, T.; Knorz, M.C.; Nagy, Z.Z. Central Corneal Volume and Endothelial Cell Count Following Femtosecond Laser-Assisted Refractive Cataract Surgery Compared to Conventional Phacoemulsification. *J. Refract. Surg.* **2012**, *28*, 387–391. [CrossRef] [PubMed]
89. Masket, S.; Sarayba, M.; Ignacio, T.; Fram, N. Femtosecond laser-assisted cataract incisions: Architectural stability and reproducibility. *J. Cataract Refract. Surg.* **2010**, *36*, 1048–1049. [CrossRef] [PubMed]
90. Palanker, D.V.; Blumenkranz, M.S.; Andersen, D.; Wiltberger, M.; Marcellino, G.; Gooding, P.; Angeley, D.; Schuele, G.; Woodley, B.; Simoneau, M.; et al. Femtosecond Laser-Assisted Cataract Surgery with Integrated Optical Coherence Tomography. *Sci. Transl. Med.* **2010**, *2*, 58ra85. [CrossRef]
91. Uy, H.S.; Shah, S.; Packer, M. Comparison of Wound Sealability Between Femtosecond Laser–Constructed and Manual Clear Corneal Incisions in Patients Undergoing Cataract Surgery: A Pilot Study. *J. Refract. Surg.* **2017**, *33*, 744–748. [CrossRef]
92. Yoo, A.; Yun, S.; Kim, J.Y.; Kim, M.J.; Tchah, H. Femtosecond Laser-assisted Arcuate Keratotomy Versus Toric IOL Implantation for Correcting Astigmatism. *J. Refract. Surg.* **2015**, *31*, 574–578. [CrossRef]
93. Dick, H.B. Future Perspectives of the Femtosecond Laser in Anterior Segment Surgery. *Ophthalmologe* **2020**, *117*, 431–436. [CrossRef]

Article

Single-Cell Elasticity Measurement with an Optically Actuated Microrobot

István Grexa [1,2], Tamás Fekete [1,3], Judit Molnár [1], Kinga Molnár [1,4], Gaszton Vizsnyiczai [1], Pál Ormos [1] and Lóránd Kelemen [1,*]

1 Biological Research Centre, Temesvári krt. 62, 6726 Szeged, Hungary; grexa.istvan@brc.hu (I.G.); fekete.tamas@brc.hu (T.F.); molnartiduj@gmail.com (J.M.); molnar.kinga@brc.hu (K.M.); gaszton@brc.hu (G.V.); ormos.pal@brc.hu (P.O.)

2 Doctoral School of Interdisciplinary Medicine, University of Szeged, Korányi fasor 10, 6720 Szeged, Hungary

3 Doctoral School of Multidisciplinary Medicine, Dóm tér 9, Hungary University of Szeged, 6720 Szeged, Hungary

4 Doctoral School of Theoretical Medicine, University of Szeged, Korányi fasor 10, 6720 Szeged, Hungary

* Correspondence: lkelemen@brc.hu; Tel.: +36-62-599-600 (ext. 419)

Received: 27 July 2020; Accepted: 20 September 2020; Published: 22 September 2020

Abstract: A cell elasticity measurement method is introduced that uses polymer microtools actuated by holographic optical tweezers. The microtools were prepared with two-photon polymerization. Their shape enables the approach of the cells in any lateral direction. In the presented case, endothelial cells grown on vertical polymer walls were probed by the tools in a lateral direction. The use of specially shaped microtools prevents the target cells from photodamage that may arise during optical trapping. The position of the tools was recorded simply with video microscopy and analyzed with image processing methods. We critically compare the resulting Young's modulus values to those in the literature obtained by other methods. The application of optical tweezers extends the force range available for cell indentations measurements down to the fN regime. Our approach demonstrates a feasible alternative to the usual vertical indentation experiments.

Keywords: cell elasticity; endothelial cells; optical micromanipulation; holographic optical tweezers; two-photon polymerization; image processing

---

1. Introduction

Autonomous microrobots and microactuators have gained attention recently due to their ability to perform complex tasks on biological targets inside microfluidic environments (channels, reservoirs) without the administration of external physical tools. The targets of these manipulations include protein [1], DNA [2], their association [3] or single cells [4,5]. Furthermore, microtools have been developed to control the flow of the solvent that carries these biological objects [6,7] or to characterize their composition [8]. The complexity of microrobots spans from simple microspheres [1,9] to complex tailor-made microstructures [4,10–12], and sometimes a group of such structures is needed to perform specific tasks [13,14]. Most often, these microtools are actuated and guided by optical means, but magnetic [15,16] or acoustic [17] controls are also applied.

Since the size of these microrobots can range from sub-micrometers to a few hundreds of micrometers, they can be easily optimized for the manipulation of single cells. A broad range of tasks can be performed: cells can be actuated with the tools, which includes their simple translation or rotation either on a hard surface [4,5] or in 3D [11,18]; the tools can enhance imaging of cells [12]; the internal structure of the cells can be altered by punching holes in them with the tools [19]; and such microtools have a great potential even in performing cell-to-cell interaction experiments with high precision and selectivity.

In this work, we report on a method that uses tailor-made microtools for the mechanical characterization of single cells. We use optically actuated microtools to make nano-indentations on the cell surface, thereby determining its elastic properties. In order to measure cell membrane elasticity, one needs to realize a small indentation on it with a known radius of curvature of the indenter and a known force; the Young's modulus then can be calculated from the measured indentation and these parameters [20,21]. In the literature, there are many works reporting on the viscoelastic properties of cells measured by atomic force microscopy (AFM). While AFM can perform this task using forces typically higher than 10 pN, the great benefit of optical manipulation is that the achievable forces complete the range of AFM reaching down to even a few tenths of a pN. Optical tweezers have been applied successfully earlier to measure cells' Young's modulus by trapping microbeads of various diameters and pushing them against the cells in an axial direction [20,22–24]. These cell indentation experiments use optical forces of less than 10 pN combined with a larger contact surface radius than a typical AFM tip ($r \approx 1$ µm vs. $r \approx 10$ nm), which allows only small indentations, and consequently only smaller Young's moduli can be measured. The smaller force and larger radius of the indenter enable the optical trap-based methods to give a more precise evaluation of the elasticity of softer cells. It is an additional aspect that in the case of the large indentations of AFM, especially if it is coupled with high indentation rates, not only elasticity but also viscosity contributes to the results [20].

On the other hand, in the arrangement where the movement of the optically actuated bead is perpendicular to the surface supporting the cells and parallel to the optical axis, the measurement of bead position is somewhat less accurate. Further, in these situations, the trapping beam illuminates the bead through the cells themselves with such high intensity that it may pose a risk of photodamage on them [25–27]. Our approach aims to overcome these drawbacks: the microtool is pressed to the cell in a lateral direction, i.e., perpendicular to the optical axis and to the cell surface that allows for measuring its position and therefore the indentations more precisely, and due to the extended shape of the tool, the trapping foci are micrometers away from the living cells under study posing no risk to them. Our microtool has two functional parts: one that interacts with the optical trapping beams and one that consists of the probe that creates the indentation on the cell surface. Its optical actuation was achieved with holographic optical tweezers (HOT) able to move the structure with 6 degrees of freedom (translations and rotations) with a precision of a few tens of nanometers. The tool can be transported anywhere in a microchannel environment and its tip can be oriented towards any lateral direction so the direction of attack can be freely selected. The probe part in the presented experiments was a tip with a few hundreds of nanometers radius, but the fabrication method allows one to freely change the radius above this value. We performed the indentation experiment on adherent endothelial cells that were cultured on a hard vertical surface that is parallel to the optical axis and formed a confluent layer. Our results demonstrate that the microtool-based method provides a Young's modulus that fits in the range reported in the literature on this cell type.

2. Materials and Methods

2.1. Microtool Design and Fabrication

The microtools, shown in Figure 1, have two functional parts. The first is used to interact with the optical field and consists of four spheres, arranged at the corners of a square with a side length of 14 µm; these spheres are to be trapped with the HOT. The second is the probe part that creates the indentation on the cells surface. This part is a rod of 2 µm length, created in the plane of the spheres 14 µm away from them; we minimized the diameter of its apex for maximal sensitivity. The rods connecting the spheres and the tip formed an X-shape to minimize interference with the optical field and were slightly offset from the plane of the spheres and the tip. This offset ensured that in the recordings of the experiments these rods were out of focus, and therefore did not add extra features to the image processing when determining the precise position of the structure.

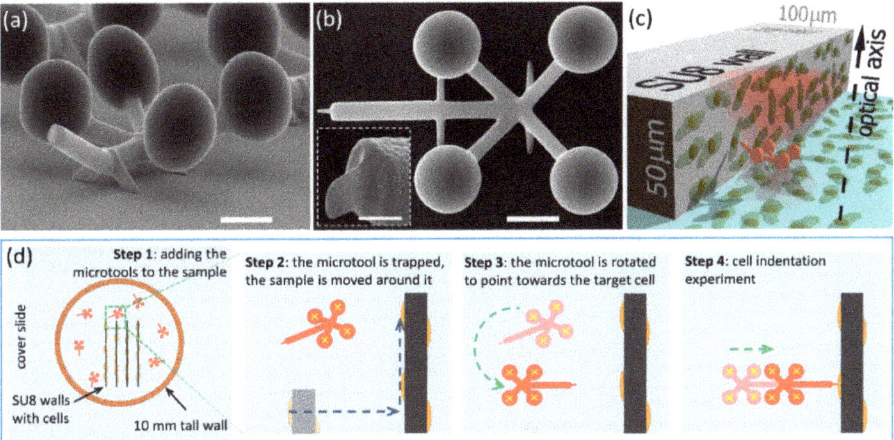

Figure 1. The polymerized microtool used for cell indentation experiments and the sample arrangement. Scanning electron microscopic images of the microtools: (**a**) side view and (**b**) top view (scale bars: 5 μm). It is visible that the tip together with the trapping spheres are at a different plane to the rods connecting them. The insert in (**b**) shows the side view of the microtool's tip (scale bar: 1 μm). (**c**) 3D schematic view of the experimental arrangement: cells are grown on a vertical wall polymerized from SU8 as well as on the glass substrate forming a confluent layer; the microtool (red structure) that is trapped and actuated with the optical tweezers (red cones) is approaching the cells on the wall with a translation that is perpendicular to the optical axis of the system. Panel (**d**) illustrates the sample assembly process with the microtools (red structures) after being pipetted into the sample well (Step 1) and their alignment towards the target cell (for details see Section 2.4); yellow crosses mark the trap beam positions, dashed blue arrows indicate sample stage movements (Step 2) and dashed green arrows the optical trap actuations (Steps 3 and 4).

The microtools were made of the photoresist SU-8 (formulation 2007) purchased from Micro Resist Technology GmbH (Berlin, Germany) together with the SU-8 developer (mr-Dev 600, Micro Resist Technology GmbH, Berlin, Germany). Their microfabrication was performed with two-photon polymerization (TPP) with the system described elsewhere [28]. Shortly, the beam of an ultrashort-pulsed laser (C-Fiber A, Menlo Systems GmbH, Martinsried, Germany, λ = 795 μm, 100 fs pulse length, 100 MHz repetition rate) was focused into a 20 μm thick photoresist layer supported by a microscope cover slide (type #1, 24 mm × 40 mm, Menzel-Glaser, TS Labor Kft, Budapest, Hungary); the focusing objective was a 100X Zeiss Achroplan, oil immersion (NA 1.25, Carl Zeiss Technika Kft, Budaörs, Hungary). The 3D scanning of the focus within the photoresist layer was carried out by a piezo stage (P-124 731.8L and P-721.10, Physik Instrumente GmbH, Karlsruhe, Germany). The illuminated SU-8 layers were processed with the standard protocol: post-exposure bake carried out at 95 °C for 10 min, development in mr-Dev 600 for 5 min 3 times, rinsing in ethanol for 5 min 3 times and finally drying with a stream of nitrogen. The microtools were removed from their support before the experiment by mechanical means in the aqueous solution of 0.5 m/m% bovine serum albumin; they were then pipetted together with this liquid and transferred to the sample containing the cells.

2.2. Cell Culturing

The cells were grown on vertical polymer surfaces (walls), which were parallel to the optical axis, as shown in Figure 1c. The walls were polymerized into SU-8 layers of about 50 μm thickness, supported by cover slides (type #1, 24 mm × 40 mm, Menzel-Glaser) using UV mask lithography. The UV light source was the 365 nm line of a mercury lamp (flood exposure source, model 97435, Newport, Irvine, CA, USA, dose: 340 mJ/cm^2). The such-created walls were ~5 mm long and 100 μm wide, positioned at the center of the cover slides. A glass ring of 10 mm height was mounted around the walls using Norland optical adhesive, thereby creating a well for cell culturing. The hCMEC/D3 human microvascular cerebral endothelial cells were grown in this well, which was tilted 45 degrees to promote cell adhesion on the vertical parts of the walls. The cells were cultured in EBM-2 medium (Lonza, Switzerland) supplemented with EGM-2 Bulletkit (Lonza, Basel, Switzerland) and 2.5% fetal bovine serum (Sigma, St. Louis, MO, USA) for 3 days before the indentation experiments. The structures, removed from their support, were pipetted in between these walls into the cell culture medium together with about 5 μL liquid that did not alter the composition of the growth medium significantly. The focused beams (red cones in Figure 1c) for the optical trapping passed into the well through its cover slide support.

2.3. HOT Setup

The cell stiffness was measured with the tailor-made microtools described above. The microtools were actuated with a holographic optical trap (HOT) system that can create multiple trapping foci and move them in 3D with high precision, as demonstrated in [12]. In the present experiments, we created 4 trapping foci forming a square of 14 μm side length and moved them with their mutual positions unchanged. The HOT system is built on an inverted Nikon microscope (Eclipse TI, Nikon, Tokyo, Japan) with a continuous-wave fiber laser (λ = 1070 nm, THFL-1P400-COL50, BKtel Photonics, Lannion, France) as a light source, an Olympus water immersion objective (UPlanSApo 60X, NA = 1.2) as a focusing element, a motorized microscope stage (ProScan, Prior Scientific, Fulbourn, UK) for sample translation and a spatial light modulator (PLUTO NIR, Holoeye, Berlin, Germany) to generate the multiple traps. The total optical power at the entrance of the objective pupil was 270 mW, which, considering the approximately 50% transmittance of the objective at 1070 nm, resulted in ~34 mW power for each trapping beam. The sample was observed with an EMCCD camera (Rolera EMC2, Qimaging, Surrey, BC, Canada).

2.4. Cell Indentation Experiment

The experimental arrangement is depicted in Figure 1c,d. Figure 1c shows the cells grown on the vertical wall polymerized onto a glass substrate, and the microtool approaching the cells in the direction which is perpendicular to the wall surface, to the cells surface and to the optical axis of the trapping objective, and parallel to the supporting glass surface. Figure 1d illustrates the sample assembly and microtool alignment procedure with the cells already present on the SU8 walls. First, the microtools, which were collected from their polymerization glass support, were pipetted into the well containing the cells (see Section 2.2). In the well, the cells were immersed in about 200–300 μL of Leibovitz's L-15 medium (Sigma) that kept them vital without CO_2 incubation for the approximately 2-h duration of the experiments. At this stage, the structures were scattered randomly on the bottom of the sample well.

For the measurement of the Young's modulus, the force that pushes the microtool to the target cell and the cell indentation need to be determined. For both values, the tool's position needs to be measured precisely. The force is calculated from the displacement of the microtool relative to the trapping foci. The indentation was determined relative to the case when it is pushed against a hard wall instead of a soft cell. The difference in the movements in these two cases provided the value of indentation, as described below. In both cases, the microtool was translated in a well-controlled manner: after taking hold of it with the optical trap, it was elevated from the substrate to about 5–15 μm above it by defocusing the objective and aligning it with the plane of its four spheres perpendicular to the optical axis. When elevated, the trap stayed fixed relative to the trapping objective and the sample stage was moved until a proper target cell arrived in the field of view (sample movement is shown by the blue dashed arrows on Figure 1d Step 2). Then, with a stationary sample stage, the microtool was rotated towards the target cell by moving the trapping focal spots, until the microtool's tip aligned with the normal of the wall's surface (with or without the cells on it) (Figure 1d Step 3 and Figure 2a,b).

After the microtool was oriented towards the cell, selected upon visual inspection, the cell's silhouette was brought into focus together with the tool's tip by the consecutive adjustments of the focusing objective and the trap position along the optical axis; this ensured that the point of attack on the cell was seen as a sharp contour. Then, the microtool approached the wall to about 2 µm, moving only the microscope stage (rough approach), and stopped. In this moment, the tool was situated at about 5–15 µm height from the supporting glass, which ensured that the trapping foci were only minimally, or not at all distorted by the bottom area of the cell-supporting wall. In step 4, the indentation experiment commenced, when the optical trap was moved in 10 nm steps towards the wall, using only the HOT (fine approach), and at each position a bright-field image of the microtool was recorded; this process resulted in an average speed of 0.05 µm/s. The trajectory of the tip of the microtool was parallel to the normal of the wall. Before making contact with the wall or the cell, the position of the microtool and the trap coincided. When the microtool reached the wall, the movement of the trapping foci continued as before for about an extra 1 µm, but the tool was retarded relative to the trap position. This retardation provides a force by which the microtool is pushed against the wall or the cell surface. Finally, the microtool was retracted and positioned to the next available cell. When more than one indentation experiment was carried out on the same cell, we probed the cell at points that were a few hundreds of nanometers away from each other. Altogether, 19 measurements were carried out on 6 cells with 4 manipulators.

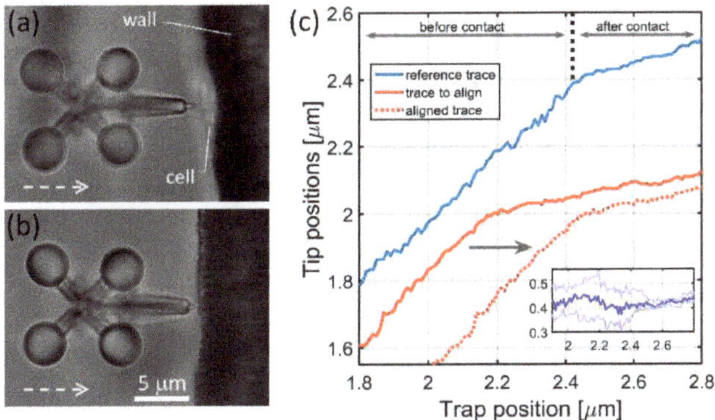

Figure 2. Cell indentation experiments and the resulted traces of the microtool's tip. Panels (**a**,**b**) show a typical snapshot of the cell indentation and the wall approaching experiments, respectively; the optical axis is perpendicular to the plane of the figure, the white dashed arrows indicate the direction of the microtool movement during the indentation experiment. The tip position was calculated by determining the positions of the four handle spheres on the image series taken during the indentation experiments. (**c**) shows tip positions from two cell indentation experiments as the function of the trapping beam position (solid blue and solid red traces). It also shows the result of the trace alignment procedure when the red trace is aligned to the blue one with the alignment procedure (dashed red). The inset in (**c**) shows differences of the red and blue traces during the alignment procedure (see main text).

2.5. Data Analysis

The position of each microtool during the approach of the cell was determined with a correlation-based method where the reference was its image at the very first position. The script was implemented in Matlab, using built-in image processing and 2D cross-correlation functions. The positions of the four trapping spheres were determined independently. First, a template image was chosen, which was the cropped image of the selected sphere on the very first frame. Then, this reference image was compared to all consecutive frames using Matlab's built-in *normxcorr2* function. This function resulted in a correlation matrix for each frame with the same size as the frame itself. The maximum of this matrix provides the position on each frame where its similarity to the reference image is the largest; in other words, this maximum is the new position of the sphere on the frame. One must pay attention to the fact that *normxcorr2* provides this position only to one pixel size precision, which was 120 nm in our case. In order to find the position with sub-pixel precision, the values around the correlation matrix maximum value were fitted with a 2D Gaussian function, and the center of this Gaussian gave the new location of the sphere with sub-pixel precision. Next, the position of the microtool's tip was calculated from these sphere position data taking advantage of the fact that the tool is a rigid structure and that it moves in the focal plane. The reason why the image of the tip itself was not monitored is that after it makes a direct contact with the cell, the image becomes distorted and it cannot be used for cross-correlation. At the end of this process, the position of the tip was determined for all the frames and could be plotted as the function of the trap position, which changes between frames by 10 nm.

The precision of the correlation-based position determination method was found to be 5.5 nm FWHM as measured on surface-attached, non-moving microtools. For this, 2000 frames of the surface-attached microtool was recorded and analyzed with the correlation-based method; in theory, the measured positions of the four spheres should not change between frames. In reality, small fluctuations were measured partly due to residual mechanical vibrations and to the imprecision of the calculation of the correlation. The positions of the tip were determined along as well as perpendicular to the direction of the optical trap movement; only those attempts were eventually used in the analysis where the tip movement perpendicular to this direction is negligible (smaller than 50 nm) after the contact.

The result of the image processing is a microtool tip position vs. trapping focus position trace for each indentation experiment. These traces have two distinct ranges as shown in Figure 2c: the first one describes the movement before the contact the microtool makes with its target; in this range, the microtool follows the trap position precisely, so its slope is 1 (reference trace before 2.42 µm in Figure 2c). After the microtool makes contact with its target (a cell, a wall or a bead, see below), it lags behind the trap, so the slope of this range becomes less than 1 (reference trace after 2.42 µm in Figure 2c). Since the contact point did not fall to the same trap position in the consecutive experiments, the traces needed to be aligned. The cell indentation experiment series and the wall approach experiments resulted in two distinct sets of traces. The cell indentation traces were aligned to each other with one alignment procedure, so were the wall approach traces with a separate procedure. In each procedure, a reference trace was selected from the experiments (usually the first one), and the rest of the traces were aligned to it. The alignment was based on calculating the variance of the difference of two traces while one of them (red curve in Figure 2c) was shifted stepwise in respect to the other one that served as a reference (blue curve in Figure 2c) (the step size was 10 nm). The minimum of the calculated variances gave the amount of trace shift used for alignment. The inset in Figure 2c shows three of such difference traces: the dark blue is the case of minimum variance, while the other two have a variance 3 times larger. After aligning the cell and wall approach experiments, the traces from the wall experiments were averaged ($n = 9$), while those of the cell experiments were used individually later to calculate cell indentation and the displacement of the microtool.

2.6. Trap Stiffness Calibration

For the calculation of the Young's modulus, the force that the microtool exerted on the cell creating the measured indentation must be known. This force is calculated as the displacement of the microtool from the equilibrium position multiplied by the trap stiffness. The microtool's trap stiffness (k_m) was measured with an indirect method. Here, the microtool was pushed against a trapped 9 µm polystyrene bead of known trap stiffness (k_b) and the displacement of the microtool was compared to that of the bead (Figure 3). First, k_b was determined by trapping the bead alone using the equation

$$\frac{1}{2}k_B T = \frac{1}{2}k_b\langle x^2\rangle \tag{1}$$

where $\langle x^2 \rangle$ is the variance of the bead fluctuation determined by video tracking (using 0.5 ms exposure time), T is room temperature (295K) and k_B is Boltzmann's constant [29]. Then, the microtool was pushed against the trapped bead and the following equation resulted in the trap stiffness of the microtool:

$$k_m = k_b \times \frac{\Delta x_b}{\Delta x_m} \tag{2}$$

where Δx_b and Δx_m are the displacements of the bead and the microtool, respectively. For this measurement, the tip of the microtool was slightly modified: instead of a sharp tip, it had a flat one; this modification was micrometers away from the trapped spheroids of the tool, so it did not affect the trap stiffness.

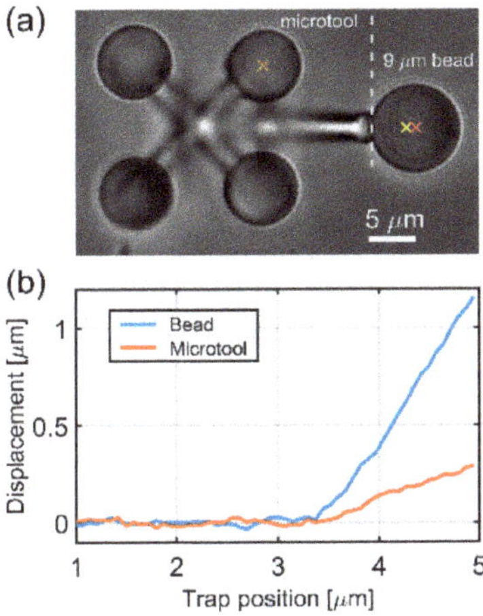

Figure 3. Trap stiffness calibration for the cell indenter microtool. Panel (**a**) shows the optical microscopic image of the tool (**left**) and the 9 μm bead (**right**) during the calibration experiment. The yellow crosses show the positions of two optical traps, one holding one of the spheroids of the microtool, the other holding the 9 μm bead. The red crosses show the center of one of the spheroids on the microtool and that of the bead. The distance of the yellow and red crosses gives the displacement values plotted on (**b**).

3. Results

An example for the polymerized microtools we used for the cell indentation experiments is shown in Figure 1. The four spheres used to hold the tool with the optical tweezers have 6.5 μm diameter and their centers form a 14 μm × 14 μm square. The apex of its tip has an ellipsoid shape with the smallest radius of 100 nm and a large one of 500 nm; this shape is inherited from the inherent shape of the basic building block of TPP. In our calculations, we take an average radius of 300 nm for the microtool tip. For the trap stiffness measurement, the tip was modified to a 2 μm × 2 μm flat end.

3.1. Trap Stiffness Calibration

In Figure 3a, the tip-modified microtool and a 9 μm bead are shown during the stiffness measurement. When the microtool is pushed against the bead in 50 nm steps, both the microtool and the bead are displaced from their equilibrium positions. Since the force that the microtool exerts on the bead and that the bead exerts on the microtool is equal, Equation (2) can be used to calculate k_m. Equation (1) resulted in a k_b bead stiffness of 4.5 pN/μm. Figure 3b depicts the averaged bead and microtool displacements as the function of the trap position. From these curves, k_m can be calculated for a range that starts about 0.5 μm after the contact point (between 4 and 5 μm); the obtained microtool stiffness value is 16.49 ± 2 pN/μm.

3.2. Endothelial Cells Young's Modulus

The Young's modulus was calculated according to the equation used in the literature for indentation experiments performed with AFM or optical tweezers (Hertz model):

$$F(d_z) = \frac{4E}{3(1-\nu^2)} R_b^{1/2} d_z^{3/2}, \tag{3}$$

where F is the force at which the indenter is pushed against the cell, E is the Young's modulus, R_b is the radius of the indenter surface, d_z is the indentation and v is the Poisson number (we chose 0.5) [21,22]. The indentations and the forces (in the form of displacement) were calculated from the microtool position traces as the function of the trap positions. Figure 4a shows representative raw microtool tip positions during the cell indentation experiments before their alignment. The individual traces illustrate that the microtools' movement changes radically after they made contact with the cells: they do not follow the movement of the trap but do not stop completely either. After the contact, the movement continues to be primarily a linear function of the trap position for at least another 500–800 nm of trap movement; in this regime, the tip moves less than 150 nm. In a few cases, the tip position suddenly increased after about 100 nm tip travel due to an occasional sideway slip on the cell membrane. Figure 4b shows one aligned tip position trace when pushed against a cell (green curve) and the average of the traces from the approaches of the hard SU8 walls (red curve). The tip position is meaningful mainly in the first 400 nm beyond the contact point, where only negligible slipping took place. In the case of the hard wall, the tips usually do not stop completely but a residual forward movement remains, which is due to the small sideway movements of the tip on the surface. We believe that these small slips also took place for the cell experiments, so the extra average forward movement observed at the walls was used as a "baseline" in the cell indentation experiments: the tip positions from the cell experiments were compared to this baseline. Figure 4c shows a typical experimental result of an approach of the hard wall. It is visible that the tip continued to move forward about 20 nm during the first 400 nm of trap position movement after the contact (between 2.43 and 2.83 μm), while it slipped sideways at an average of 50 nm.

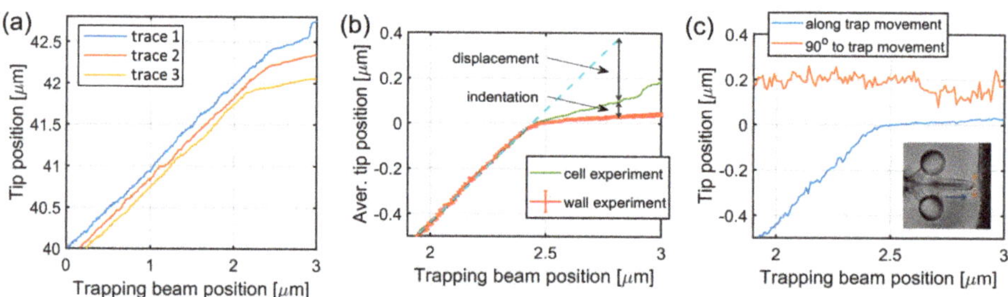

Figure 4. Tip position traces for the calculation of endothelial cell's Young's modulus. Panel (**a**) shows representative raw tip position traces as the function of the trap positions, before alignment, obtained from individual cell indentation experiments. After the alignment procedure, the tip positions (**b**) of the wall approach experiments were averaged for background (red curve), while the cell indentation traces (an example is shown by the green curve) were used individually to calculate indentation and displacement. The error bars on the red trace represent standard deviation. At each trapping beam position, the cell indentation was calculated as the difference between the green and red curves, and the microtool displacement as the difference between the green curve and the trapping beam position (dashed light blue curve). Panel (**c**) shows a tip position movement parallel (blue) and perpendicular (red) to the trap movement (that is, axis of the microtool) during the tool being pushed against a hard wall. The inset displays the tip movement along the trap progression (blue line) and perpendicular to it (red line).

The two position traces illustrated in Figure 4b were used to calculate the indentation and force values used in the Hertz model for each individual cell indentation experiment. The indentation is simply the difference between the tip positions when approaching the cell and when approaching the wall. The displacement was calculated by first fitting a straight line to the initial part of the cell approach trace (light blue dashed line in Figure 4b) and then taking the difference between this line and the tip position after the contact point. Figure 5a shows these two values, the cell indentation and the microtool displacement as the function of the trap position. The applied force is calculated from the displacement by multiplying it by the trap stiffness k_m. The displacement resulted in a force ranging from 1 to 5 pN, which is below the precision of an AFM. Both traces have a break at the contact point between 2.4 and 2.5 µm, but produce a large error of about 400 nm after the contact point. The displacement and indentation values can be measured reliably in the trap position range of 2.5–2.9 µm, therefore the Young's modulus can be regarded as reliable also only here. We obtained values ranging from about 220 up to about 1500 Pa (Figure 5b), although between the 2.5 and 2.6 µm trap positions (corresponding indentation: between 0.01 and 0.02 µm), the values could be determined with significant noise.

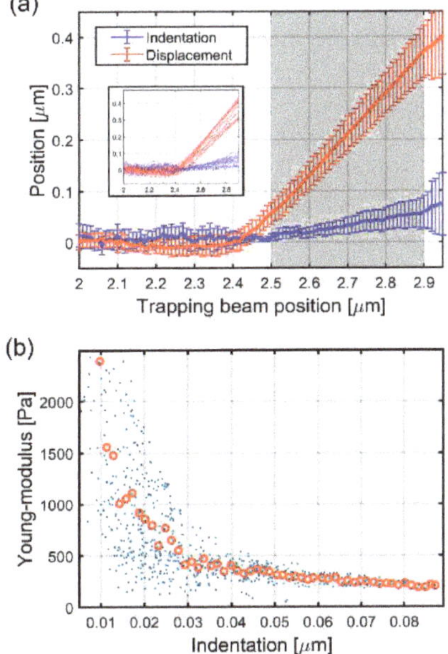

Figure 5. The measured indentation and displacement data and the Young's modulus calculated from them. The indentation (blue line) and displacement (red line) data on panel (**a**) are calculated from the aligned traces of the 19 cell indentation experiments as shown in Figure 4b; the error bars represent standard deviation. The shaded area highlights the reliable range for the two quantities. The inset shows the individual displacement traces (red) and indentation traces (blue) calculated separately for the 19 experiments. The Young's modulus as the function of indentation over the values highlighted in (**a**) is shown in panel (**b**). The blue dots represent all of the approximately 800 individual point pairs (40 trap positions × 19 experiments), while the red circles are their averages in 60 regions over the 0–0.09 µm indentation range.

4. Discussion

We have designed an optical tweezers-operated microtool specifically to measure the elasticity of adherent cells in closed microfluidic environments. In contrast to earlier optical trap-based cell indentation experiments, this tool approaches the target cells laterally which makes the determination of its position easier even with simple bright-field video microscopy. In addition, in the trapped parts of the structures, the four spheroids are more than ten micrometers away from the probe tip, which ensures that the optical field does not cause photodamage to the cells. The analysis of the microtools' position during the indentation experiments revealed that they move in a straight and continuous way after making contact with the cell, and their position values yield the indentation and pushing force values in a straightforward manner. The sideways slip of the microtools' tip in contact with the cells could be readily detected and excluded from the evaluation based on its magnitude. Residual sideways tip movements could be compensated for with a control measurement using a hard wall without cells. The microtools were characterized with 16.5 pN/µm trap stiffness and pushing force in the 1–5 pN range, comparable to other optical trap-based elasticity measurements but well below that of an AFM. This force together with the 300 nm tip radius of the microtool yielded the measured indentation values of up to 90 nm. The operating force range can be easily extended to higher values with higher trapping laser power and smaller trapping sphere diameters.

We obtained Young's moduli in the range between 220 and 1500 Pa, depending upon the position of the trapping beam, consequently on the amount of indentation; the noise of the Young's modulus, however, remarkably increases with the decrease in the indentation. These values are comparable to those measured by AFM on bovine aortic endothelial cells (700–2.7 kPa [30]), on pulmonary artery endothelial cells (400–1500 Pa [31]), on human umbilical vein endothelial cells (HUVEC) (350–4000 Pa [21]) or by magnetic tweezers on HUVEC (400 Pa [32]). It is noticeable that the values for the AFM-derived Young's modulus can easily vary an order of magnitude across the literature for the same type of cell. For endothelial cells, values anywhere from ~200 to ~5000 Pa can be found; our measured Young's modulus falls to the lower regime with a few hundred Pa. The main reasons for this broad range can be found in the measurement conditions: mainly in the rate and amount of indentation, and in the shape of the object the indentation is realized with.

It was shown that increasing the indentation rate increases the apparent Young's modulus due primarily to viscous effects [20,33]. The typical loading rates used in an AFM measurement span from 100 pN/s [20] to tens [33] or hundreds of nN/s values (exerting 1 nN force with 0.5 kHz frequency of the AFM cantilever [34]). Our optical tweezers approach experiment lasts about 60 s, where during only the last 5–6 s does the tip actually hit the cell. Considering the averaged maximum of 60 nm indentation and that by the end of this period the force increases to an average of 6 pN, it yields an indentation rate as low as 0.01 µm/s and a loading rate of about 1 pN/s, which is orders of magnitude smaller than those of AFM. For Mathur and co-workers, the lower limit for viscous dissipation was at 0.25 µm/s probe velocity [33]. We are confident that at the observed low-indentation rate viscous effects do not play any role in measuring the Young's modulus in our experiments.

In the papers of Vargas-Pinto and Mathur [21,33], the authors also showed that the higher the indentation, the lower the Young's modulus, similarly to our results (Figure 5b). The amount of indentation, which for us was up to 90 nm, is in the lower regime of what was obtained with AFM or optical tweezers [21,24]; it is very likely that with this low value, we mainly measure the elastic properties of the cell membrane and not the complex characteristics of the underlying actin network. The measurement error of Young's modulus increases significantly below 30 nm indentation. It is believed that this noise is primarily due to thermal fluctuations and the increase in the relative error when using small indentation and force values in Equation (3). However, the larger E for lower indentations is elsewhere argued to originate from the nonlinear elasticity of the cell [33]; it is beyond the scope of this paper to study the nonlinear phenomenon in detail. The third important parameter is the shape of the intender. Vargas-Pinto et al. reported on using AFM tips with radii of curvature from 20 nm to 5 µm for endothelial cell indentation [21]. They found that while the sharp tip yielded a value of 3.8 kPa for the Young's modulus, the 5 µm one yielded only 350 Pa for the same type of cell. Similarly, Harris and co-workers found that pyramidal sharp tips can measure double the Young's modulus of that measured with spherical (r = 7.5 µm) tips (800 vs. 400 Pa) on MDCK cells [35]. Chiou and co-workers also observed a more than two-fold increase in the Young's modulus value for mouse fibroblast cells when they compared sharp pyramidal tips with flat top (diameter of 1.8 μm) and spherical (sphere r = 2.5 μm) tips [36]. The tip of our microtool has a radius of 300 nm on average, which is much larger than that of the conical AFM ($r \approx$ 10 nm) tips and comparable to those used for optical tweezers indentation (r = 0.4–1.5 μm); this size also points towards measuring Young's moduli in the lower few hundreds of the Pa regime with our microtool.

In conclusion, the measurement of a cell's Young's modulus requires a careful approach in order to obtain reliable results. Even with one technique, one can measure very different values depending on the measurement parameters. The solution probably lies in what one actually wants to measure. If one is interested in the pure linear elastic properties of the cell, it is believed that the use of large radius of curvature indentation surfaces, small indentations (with small forces) and small loading rates is more appropriate to characterize specifically that. Optically micro-manipulated polymer structures should ideally operate in this regime.

Author Contributions: I.G. and L.K. conceived and designed the experiments; I.G., T.F., K.M. and J.M. performed the experiments; I.G., G.V. and L.K. designed the data analysis software; I.G. and L.K. wrote the manuscript; G.V. and P.O. contributed to the interpretation of the results; P.O. funded the research work. All authors have read and agreed to the published version of the manuscript.

Funding: This work was supported by the GINOP-2.3.2-15- 2016-00001 and the GINOP-2.3.3-15-2016-00040 programs. We gratefully acknowledge the support of NVIDIA Corporation with the donation of the Titan Xp GPU used for this research. Kinga Molnár was supported by the Ministry of Innovation and Technology, Hungary (grant No.: ÚNKP-20-4-SZTE-138).

Conflicts of Interest: The authors declare no conflict of interest.

References

1. Chen, H.; Chandrasekar, S.; Sheetz, M.P.; Stossel, T.P.; Nakamura, F.; Yan, J. Mechanical perturbation of filamin A immunoglobulin repeats 20-21 reveals potential non-equilibrium mechanochemical partner binding function. *Sci. Rep.* **2013**, *3*, 1642. [CrossRef] [PubMed]
2. Oroszi, L.; Galajda, P.; Kirei, H.; Bottka, S.; Ormos, P. Direct measurement of torque in an optical trap and its application to double-strand DNA. *Phys. Rev. Lett.* **2006**, *97*, 058301. [CrossRef] [PubMed]
3. Simons, M.; Pollard, M.R.; Hughes, C.D.; Ward, A.D.; Van Houten, B.; Towrie, M.; Botchway, S.W.; Parker, A.W.; Kad, N.M. Directly interrogating single quantum dot labelled UvrA2 molecules on DNA tightropes using an optically trapped nanoprobe. *Sci. Rep.* **2015**, *5*, 18486. [CrossRef] [PubMed]
4. Aekbote, B.L.; Fekete, T.; Jacak, J.; Vizsnyiczai, G.; Ormos, P.; Kelemen, L. Surface-modified complex SU-8 microstructures for indirect optical manipulation of single cells. *Biomed. Opt. Express* **2016**, *7*, 45–56. [CrossRef] [PubMed]
5. Hu, S.; Xie, H.; Wei, T.; Chen, S.; Sun, D. Automated Indirect Transportation of Biological Cells with Optical Tweezers and a 3D Printed Microtool. *Appl. Sci.* **2019**, *9*, 2883. [CrossRef]
6. Maruo, S.; Takaura, A.; Saito, Y. Optically driven micropump with a twin spiral microrotor. *Opt. Express* **2009**, *17*, 18525–18532. [CrossRef]
7. Terray, A.; Oakey, J.; Marr, D.W.M. Microfluidic Control Using Colloidal Devices. *Science* **2002**, *296*, 1841–1844. [CrossRef]
8. Vizsnyiczai, G.; Lestyan, T.; Joniova, J.; Aekbote, B.L.; Strejckova, A.; Ormos, P.; Miskovsky, P.; Kelemen, L.; Bano, G. Optically Trapped Surface-Enhanced Raman Probes Prepared by Silver Photoreduction to 3D Microstructures. *Langmuir* **2015**, *31*, 10087–10093. [CrossRef]
9. Mártonfalvi, Z.; Bianco, P.; Naftz, K.; Ferenczy, G.G.; Kellermayer, M. Force generation by titin folding. *Protein Sci.* **2017**, *26*, 1380–1390. [CrossRef]
10. Barbot, A.; Decanini, D.; Hwang, G. On-chip Microfluidic Multimodal Swimmer toward 3D Navigation. *Sci. Rep.* **2016**, *6*, 19041. [CrossRef]
11. Vizsnyiczai, G.; Aekbote, B.L.; Buzas, A.; Grexa, I.; Ormos, P.; Kelemen, L. High accuracy indirect optical manipulation of live cells with functionalized microtools. In *Optical Trapping and Optical Micromanipulation Xiii*; Dholakia, K., Spalding, G.C., Eds.; SPIE: Bellingham, WA, USA, 2016; Volume 9922, p. 992216.

12. Vizsnyiczai, G.; Búzás, A.; Lakshmanrao Aekbote, B.; Fekete, T.; Grexa, I.; Ormos, P.; Kelemen, L. Multiview microscopy of single cells through microstructure-based indirect optical manipulation. *Biomed. Opt. Express* **2020**, *11*, 945–962. [CrossRef] [PubMed]
13. Būtaitė, U.G.; Gibson, G.M.; Ho, Y.-L.D.; Taverne, M.; Taylor, J.M.; Phillips, D.B. Indirect optical trapping using light driven micro-rotors for reconfigurable hydrodynamic manipulation. *Nat. Commun.* **2019**, *10*, 1215. [CrossRef] [PubMed]
14. Turlier, H.; Fedosov, D.A.; Audoly, B.; Auth, T.; Gov, N.S.; Sykes, C.; Joanny, J.F.; Gompper, G.; Betz, T. Equilibrium physics breakdown reveals the active nature of red blood cell flickering. *Nat. Phys.* **2016**, *12*, 513. [CrossRef]
15. Bente, K.; Codutti, A.; Bachmann, F.; Faivre, D. Biohybrid and Bioinspired Magnetic Microswimmers. *Small* **2018**, *14*, 1704374. [CrossRef] [PubMed]
16. Sarkar, R.; Rybenkov, V.V. A Guide to Magnetic Tweezers and Their Applications. *Front. Phys.* **2016**, *4*. [CrossRef]
17. Sitters, G.; Kamsma, D.; Thalhammer, G.; Ritsch-Marte, M.; Peterman, E.J.G.; Wuite, G.J.L. Acoustic force spectroscopy. *Nat. Methods* **2015**, *12*, 47–50. [CrossRef]
18. Shishkin, I.; Markovich, H.; Roichman, Y.; Ginzburg, P. Auxiliary Optomechanical Tools for 3D Cell Manipulation. *Micromachines* **2020**, *11*, 90. [CrossRef]
19. Hayakawa, T.; Fukada, S.; Arai, F. Fabrication of an On-Chip Nanorobot Integrating Functional Nanomaterials for Single-Cell Punctures. *IEEE Trans. Robot.* **2014**, *30*, 59–67. [CrossRef]
20. Nawaz, S.; Sanchez, P.; Bodensiek, K.; Li, S.; Simons, M.; Schaap, I.A.T. Cell Visco-Elasticity Measured with AFM and Optical Trapping at Sub-Micrometer Deformations. *PLoS ONE* **2012**, *7*, e45297. [CrossRef]
21. Vargas-Pinto, R.; Gong, H.; Vahabikashi, A.; Johnson, M. The Effect of the Endothelial Cell Cortex on Atomic Force Microscopy Measurements. *Biophys. J.* **2013**, *105*, 300–309. [CrossRef]
22. Dy, M.C.; Kanaya, S.; Sugiura, T. Localized cell stiffness measurement using axial movement of an optically trapped microparticle. *J. Biomed. Opt.* **2013**, *18*. [CrossRef] [PubMed]
23. Ndoye, F.; Yousafzai, M.S.; Coceano, G.; Bonin, S.; Scoles, G.; Ka, O.; Niemela, J.; Cojoc, D. The influence of lateral forces on the cell stiffness measurement by optical tweezers vertical indentation. *Int. J. Optomechatron.* **2016**, *10*, 53–62. [CrossRef]
24. Yousafzai, M.S.; Coceano, G.; Mariutti, A.; Ndoye, F.; Amin, L.; Niemela, J.; Bonin, S.; Scoles, G.; Cojoc, D. Effect of neighboring cells on cell stiffness measured by optical tweezers indentation. *J. Biomed. Opt.* **2016**, *21*, 57004. [CrossRef] [PubMed]
25. Ayano, S.; Wakamoto, Y.; Yamashita, S.; Yasuda, K. Quantitative measurement of damage caused by 1064-nm wavelength optical trapping of Escherichia coli cells using on-chip single cell cultivation system. *Biochem. Biophys. Res. Commun.* **2006**, *350*, 678–684. [CrossRef] [PubMed]
26. Liang, H.; Vu, K.T.; Krishnan, P.; Trang, T.C.; Shin, D.; Kimel, S.; Berns, M.W. Wavelength dependence of cell cloning efficiency after optical trapping. *Biophys. J.* **1996**, *70*, 1529–1533. [CrossRef]
27. Rasmussen, M.B.; Oddershede, L.B.; Siegumfeldt, H. Optical tweezers cause physiological damage to Escherichia coli and Listeria bacteria. *Appl. Environ. Microbiol.* **2008**, *74*, 2441–2446. [CrossRef] [PubMed]
28. Vizsnyiczai, G.; Kelemen, L.; Ormos, P. Holographic multi-focus 3D two-photon polymerization with real-time calculated holograms. *Opt. Express* **2014**, *22*, 24217–24223. [CrossRef] [PubMed]
29. Neuman, K.C.; Block, S.M. Optical trapping. *Rev. Sci. Instrum.* **2004**, *75*, 2787–2809. [CrossRef]
30. Ohashi, T.; Ishii, Y.; Ishikawa, Y.; Matsumoto, T.; Sato, M. Experimental and numerical analyses of local mechanical properties measured by atomic force microscopy for sheared endothelial cells. *Bio-Med. Mater. Eng.* **2002**, *12*, 319–327.
31. Pesen, D.; Hoh, J.H. Micromechanical Architecture of the Endothelial Cell Cortex. *Biophys. J.* **2005**, *88*, 670–679. [CrossRef]
32. Feneberg, W.; Aepfelbacher, M.; Sackmann, E. Microviscoelasticity of the Apical Cell Surface of Human Umbilical Vein Endothelial Cells (HUVEC) within Confluent Monolayers. *Biophys. J.* **2004**, *87*, 1338–1350. [CrossRef] [PubMed]
33. Mathur, A.B.; Collinsworth, A.M.; Reichert, W.M.; Kraus, W.E.; Truskey, G.A. Endothelial, cardiac muscle and skeletal muscle exhibit different viscous and elastic properties as determined by atomic force microscopy. *J. Biomech.* **2001**, *34*, 1545–1553. [CrossRef]

34. Coceano, G.; Yousafzai, M.S.; Ma, W.; Ndoye, F.; Venturelli, L.; Hussain, I.; Bonin, S.; Niemela, J.; Scoles, G.; Cojoc, D.; et al. Investigation into local cell mechanics by atomic force microscopy mapping and optical tweezer vertical indentation. *Nanotechnology* **2016**, *27*, 065102. [CrossRef] [PubMed]
35. Harris, A.R.; Charras, G.T. Experimental validation of atomic force microscopy-based cell elasticity measurements. *Nanotechnology* **2011**, *22*, 345102. [CrossRef]
36. Chiou, Y.W.; Lin, H.K.; Tang, M.J.; Lin, H.H.; Yeh, M.L. The Influence of Physical and Physiological Cues on Atomic Force Microscopy-Based Cell Stiffness Assessment. *PLoS ONE* **2013**, *8*. [CrossRef]

Article

Femtosecond Laser Direct Writing of Integrated Photonic Quantum Chips for Generating Path-Encoded Bell States

Meng Li [1], Qian Zhang [1], Yang Chen [2], Xifeng Ren [2], Qihuang Gong [1,3,4] and Yan Li [1,3,4,*]

1. State Key Laboratory for Mesoscopic Physics and Frontiers Science Center for Nano-Optoelectronics, School of Physics, Peking University, Beijing 100871, China; mengli2016@pku.edu.cn (M.L.); zhangqianlucy@126.com (Q.Z.); qhgong@pku.edu.cn (Q.G.)
2. CAS Key Laboratory of Quantum Information, University of Science and Technology of China, Hefei 230026, China; cy123@mail.ustc.edu.cn (Y.C.); renxf@ustc.edu.cn (X.R.)
3. Collaborative Innovation Center of Extreme Optics, Shanxi University, Taiyuan 030006, China
4. Peking University Yangtze Delta Institute of Optoelectronics, Nantong 226010, China

* Correspondence: li@pku.edu.cn

Received: 27 November 2020; Accepted: 13 December 2020; Published: 15 December 2020

Abstract: Integrated photonic quantum chip provides a promising platform to perform quantum computation, quantum simulation, quantum metrology and quantum communication. Femtosecond laser direct writing (FLDW) is a potential technique to fabricate various integrated photonic quantum chips in glass. Several quantum logic gates fabricated by FLDW have been reported, such as polarization and path encoded quantum controlled-NOT (CNOT) gates. By combining several single qubit gates and two qubit gates, the quantum circuit can realize different functions, such as generating quantum entangled states and performing quantum computation algorithms. Here we demonstrate the FLDW of integrated photonic quantum chips composed of one Hadamard gate and one CNOT gate for generating all four path-encoded Bell states. The experimental results show that the average fidelity of the reconstructed truth table reaches as high as 98.8 ± 0.3%. Our work is of great importance to be widely applied in many quantum circuits, therefore this technique would offer great potential to fabricate more complex circuits to realize more advanced functions.

Keywords: photonic quantum chip; femtosecond laser direct writing; Hadamard gate; CNOT gate; path-encoded Bell state

---

1. Introduction

In recent years, integrated photonic quantum chips have become a hot topic in quantum optics field, for its scalability, stability, and miniaturization compared with bulk optics. They are fabricated by the silicon-based lithography [1–6], the femtosecond laser direct writing (FLDW) [7–9], and a new emerging platform based on lithium niobate on insulator (LNOI) [10–12]. Silicon-based waveguide photonic chips have maturely developed using silica on silicon [1], silicon on insulator [4], silicon nitride [5], silicon oxynitride [6] and so on, while their waveguides can only support single polarization mode due to the rectangular cross-section, and the conventional lithography is limited to the planar layout. The FLDW can realize 3D fabrication of waveguides with near round cross-section [13–15]. Therefore, it can realize quantum information processing not only by path encoding [8,16] but also by polarization encoding [17,18]. By virtue of its true three-dimensional direct writing ability, FLDW can fabricate more flexible and complex quantum circuits with 3D structures, simplify the layout and reduce the number of elements required [19–21]. Nowadays, photonic quantum chips realized by FLDW technique have been widely applied in various research, such as quantum logic gates [8,17], quantum algorithm [22–24], quantum walk [25–27], quantum simulation [28–30], quantum key distribution [31,32], boson sampling [33,34], entangled photon sources [35], and so on.

Quantum logic gates are the basic elements of quantum circuits. Combining several simple logic gates together can construct more complex logic gates to perform quantum algorithms and realize specific functions [36–39]. The Hadamard (H) gate and the Controlled-NOT (CNOT) gate are the most basic and important single and two qubit gates, respectively. Cascading the H gate and the CNOT gate together can generate Bell states, which are also the most basic entangled states and serve as a central physical resource in various quantum information protocols like quantum cryptography, quantum teleportation, entanglement swapping, and in tests aimed at excluding hidden variable models of quantum mechanics [40]. Up to now, individual path-encoded or polarization-encoded H gate or CNOT gate on chips has already been fabricated by FLDW technique [16–18,41]. However, the combination of one H gate and one CNOT gate in a single photonic chip for generating path-encoded Bell states has not been reported. Here we have successfully fabricated an integrated photonic quantum chip composed of H and CNOT gates by FLDW technique to generate all four path-encoded Bell entangled states, whose average fidelity is 98.8 ± 0.3%, which is higher than that of a silicon-based photonic chip (~91.2 ± 0.2%) [1]. This combination is very useful in many quantum circuits, especially in quantum algorithms circuits [2,19,42]. The capacity of FLDW to fabricate such a chip with high quality and fidelity provides a possibility for the fabrication of large-scale 3D functional integrated photonic quantum chips.

2. Design of the Photonic Quantum Chip

Figure 1a shows the schematic configuration of the photonic quantum chip composed of one H gate and one CNOT gate. The H gate (red dashed box) is a balanced directional coupler (DC), and the CNOT gate (the rest of the whole chip) is based on our previous work [16] but with improved symmetry in circuit design. The left represents the input ports (0–5) and the right the output ports (0′–5′). The power reflectivity $R = P_{\text{OUT1}}/(P_{\text{OUT1}} + P_{\text{OUT2}})$ is marked on each DC, defined by the ratio of the output power from OUT1 to the total output power of a DC where the laser is launched into IN1 as shown in Figure 1b. The symmetry of the DC guarantees that the same relations hold when light is launched into port IN2, by simply inverting the two indices. There are three DCs with reflectivity of 1/2 and three DCs with reflectivity of 1/3. The control qubit C_{q} (target qubit T_{q}) is encoded via spatial paths $C_0(T_0)$ and $C_1(T_1)$. The remaining two paths are ancillary vacuum modes to complete the network. The H gate in this circuit represented by a DC is indeed a Hadamard-like gate $H' = e^{i\pi/2}e^{-i\pi\sigma_z/4}He^{-i\pi\sigma_z/4} = \frac{1}{\sqrt{2}}\begin{pmatrix} 1 & i \\ i & 1 \end{pmatrix}$ as shown in Reference [3], where $H = \frac{1}{\sqrt{2}}\begin{pmatrix} 1 & 1 \\ 1 & -1 \end{pmatrix}$ is the standard Hadamard gate. They are equivalent up to local σ_z rotations, so we still use H gate for description in this article. When a single photon is input to path $C_0(C_1)$ representing the logic state $|0\rangle_c(|1\rangle_c)$, the state will be unitarily transformed by the H gate to generate a superposition state $\frac{1}{\sqrt{2}}(|0\rangle_c + i|1\rangle_c)$ $(\frac{1}{\sqrt{2}}i(|0\rangle_c - i|1\rangle_c))$. The rest of the chip is a path encoded probabilistic CNOT gate. In the CNOT gate, when the control qubit C_q is in logic state $|1\rangle$, the target qubit T_q will flip from initial logic state $|0\rangle(|1\rangle)$ to opposite logic state $|1\rangle(|0\rangle)$. However, when C_q is in logic state $|0\rangle$, the logic state of T_q remain unchanged. When one photon from C_q and the other photon from T_q are mixed in the central DC with reflectivity of 1/3 simultaneously, they will undergo a partial bunching effect and get a π phase shift due to Hong–Ou–Mandel (HOM) interference of indistinguishable photons [16,43]; and this π phase shift will change the output state of Mach–Zehnder (MZ) interferometer connected by paths $T_0 - T_1'$ and $T_1 - T_0'$.

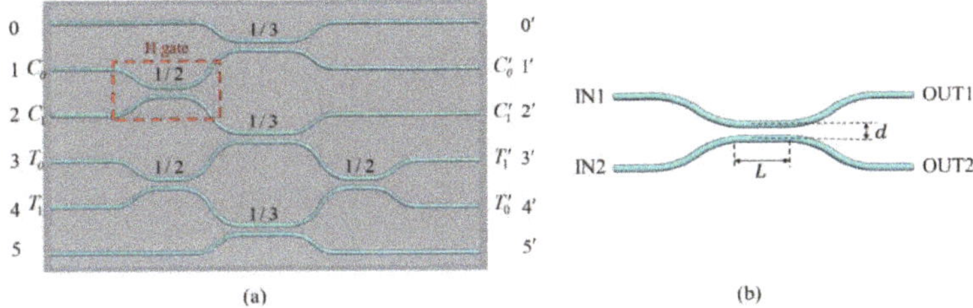

Figure 1. Schematic of the photonic quantum chip and the directional coupler (DC). (**a**) Schematic representation of a photonic chip composed of one H and one CNOT gate to generate path-encoded Bell states. The DC in the red dashed box represents an H gate. (**b**) Schematic representation of a waveguide directional coupler; IN1 and IN2 are the input ports, whereas OUT1 and OUT2 are the output ports of the device; L is the interaction length and d is the interaction distance in the coupling region of two adjacent waveguides.

Now, when the control qubit C_q is in superposition state of $|0\rangle$ and $|1\rangle$, the chip will generate all four path-encoded Bell states with the entangling function of the CNOT gate. The relationships are as follows:

$$\begin{aligned} |00\rangle_{ct} &\rightarrow \tfrac{1}{\sqrt{2}}i(|00\rangle_{ct} - |11\rangle_{ct}), \\ |01\rangle_{ct} &\rightarrow \tfrac{1}{\sqrt{2}}i(|01\rangle_{ct} + |10\rangle_{ct}), \\ |10\rangle_{ct} &\rightarrow -\tfrac{1}{\sqrt{2}}(|00\rangle_{ct} + |11\rangle_{ct}), \\ |11\rangle_{ct} &\rightarrow -\tfrac{1}{\sqrt{2}}(|01\rangle_{ct} - |10\rangle_{ct}). \end{aligned} \tag{1}$$

Compared with the standard quantum circuit composed of one H gate and one CNOT gate, the output Bell states for input state $|00\rangle_{ct}$ and $|10\rangle_{ct}$ in Equation (1) should be exchanged, because the H gate here is indeed an H′ gate.

3. Experimental and Results

The photonic quantum chip is directly written inside borosilicate glass (Eagle2000, Corning) by focusing the femtosecond laser pulses produced by a regeneratively amplified Yb: KGW femtosecond laser system (Pharos-20W-1MHz, Light Conversion). The 1030 nm laser pulses with duration of 240 fs at repetition rate of 1 MHz are focused 170 μm beneath the surface of the glass by a microscope objective with a numerical aperture (NA) of 0.5 (RMS20X-PF, Olympus). The sample is translated at constant speed executed by a computer-controlled high-precision three-axis air-bearing stage (FG1000-150-5-25-LN, Aerotech).

The first step is to fabricate straight waveguide to determine the optimal parameters by scanning pulse energy and translation speed. In this work, they are 386 nJ and 20 mm/s. Figure 2a shows the optical micrograph of the cross section of the fabricated straight waveguide, with a size of 4.5 × 7.1 μm. The mode distributions of the 785 nm laser guided in the fiber and in the waveguide with two orthogonal polarizations are shown in Figure 2b–e. The mode field diameter (MFD) of the guided mode in the waveguide is 6.1 × 6.6 μm (5.9 × 6.6 μm) in H (V) polarization with an almost round-shaped mode field. The mode in the waveguide slightly deviates from that in the fiber, which results in coupling loss due to the mode mismatching. The measured insertion loss of the 2.5 cm straight waveguide is 2.52 dB (2.43 dB), and the coupling loss is 0.80 dB (0.73 dB), and the Fresnel reflection loss is 0.177 dB/facet, so the propagation loss is 0.55 dB/cm (0.54 dB/cm) for H (V) polarization, which is better than 0.7 dB/cm in our previous work [16].

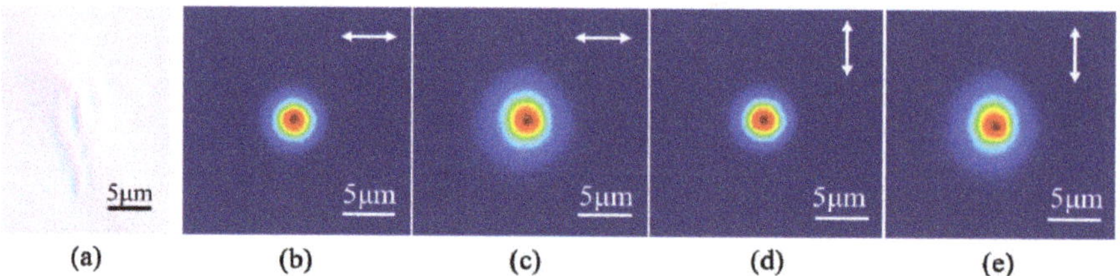

Figure 2. Optical micrograph and mode distributions of a fabricated straight waveguide. (**a**) Micrograph of the cross section of the waveguide. (**b**) Mode of the fiber in H polarization; (**c**) mode of the waveguide in H polarization; (**d**) mode of the fiber in V polarization; (**e**) mode of the waveguide in V polarization. The wavelength of the injected CW laser is 785 nm.

Using the same parameters, we fabricated a series of directional couplers with different interaction lengths L at fixed interaction distance d. For the curved segments of DC, the bending radius is set as 60 mm to guarantee a low bending loss. The spacing between two input (output) ports of the DC is 127 μm to match the pitch of the fiber array. The interaction distance is set as $d = 8$ μm to acquire a high coupling coefficient κ but without waveguide-overlapping geometrically, according to previous experimental results [16]. We change the interaction lengths in a wide range from 0–5.5 mm to find the optimal parameters. As shown in Figure 3a, the fitting curves of the measured reflectivity R and transmission T of DCs varying with interaction length L follow the curves in the forms of $\cos^2\varphi$ and $\sin^2\varphi$ very well, respectively. Therefore, we can get the linear relation between the coupling phase φ and the interaction length L: $\varphi = \kappa L + \varphi_0$, which is shown in Figure 3b, so that we can conveniently estimate the L for R = 1/3 from the data for R = 1/2. To fabricate DCs with R = 1/2 and R = 1/3, we only focus on the range of L from 0–1 mm. According to the design of the circuit in Figure 1a, we fabricated several chips composed of one H gate and one CNOT gate with slightly different interaction lengths L around the values of $L_{1/2}$ and $L_{1/3}$ to take into account possible fabrication imperfections. Finally, we successfully find satisfactory interaction lengths for DCs with R = 1/2 and R = 1/3, which are $L_{1/2}$ = 0.450 mm and $L_{1/3}$ = 0.716 mm, respectively, and the chip size is about 635 μm × 2.5 cm.

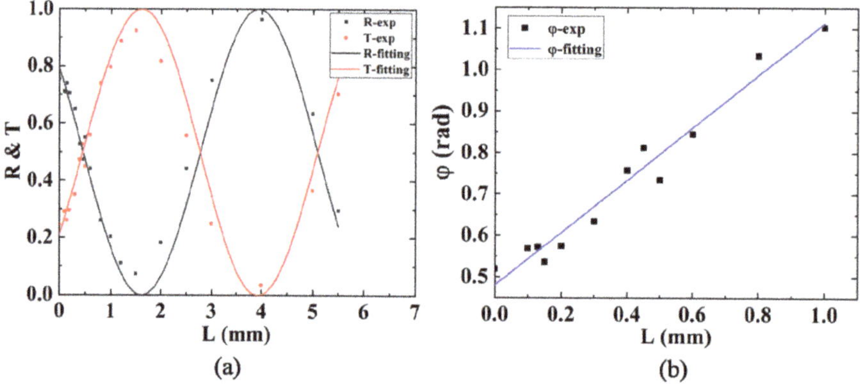

Figure 3. Experimental reflectivity and transmission of fabricated DCs. (**a**) Measured reflectivity R and transmission T of the DCs with different interaction length L at fixed interaction distance d = 8 μm as well as fitting curves for R and T. (**b**) Fitting for linear relation between coupling phase φ and interaction length L ranging from 0–1 mm.

A classical characterization of the quantum chip is performed by injecting 808 nm CW laser beam in V polarization into the chip to measure the output power of each port, we select out the best one whose results are close to the theoretical values to perform quantum characterization. In classical characterization, we input CW laser into each input port of the chip and use the power meter to record the ratio of the output power for each output port. The theoretical prediction of the ratio of output power for each input case of the circuit is listed in Table 1, and the experimental classical characterization results are shown in Table 2, where $F = \sum_i \sqrt{p_i q_i}$ (p_i is the theoretical value, and q_i is the experimental one) represents the fidelity of each row. In view of the fabrication imperfection and slightly asymmetric beam splitting ratio of fabricated DCs for different input ports, the experimental results deviate a little from the theoretical values, but they are acceptable according to the calculated fidelity of each row. Moreover, the latter quantum characterization experimental results also confirm the high quality of this chip.

Table 1. Theoretical prediction of the ratio of output power for each output port y′(0′–5′) when laser is injected into each input port x (0–5).

Theory	0′	1′	2′	3′	4′	5′
0	1/3	2/3				
1	1/3	1/6	1/6	1/6	1/6	
2	1/3	1/6	1/6	1/6	1/6	
3			1/3	0	1/3	1/3
4			1/3	1/3	0	1/3
5				1/3	1/3	1/3

Table 2. Experimental classical characterization results of the selected chip for each input case.

Experiment	0′	1′	2′	3′	4′	5′	F
0	0.350	0.650					0.999
1	0.352	0.181	0.159	0.156	0.153		0.999
2	0.355	0.180	0.155	0.155	0.155		0.999
3			0.297	0.030	0.318	0.355	0.984
4			0.301	0.306	0.028	0.365	0.985
5				0.348	0.375	0.277	0.998

For the quantum characterization of the selected chip, we inject the time-correlated photon pairs into the chip and measure the coincidence counts of output photons. The experimental setup of the quantum characterization system is shown in Figure 4. The 808 nm dual photon pairs were generated by pumping a 0.5 mm thick beta-barium borate (BBO) crystal using 140 mW, 404 nm CW laser (ECL801, Uni Quanta) through Type-I spontaneous parametric down-conversion (SPDC) process. The photon pairs are divided into two parts and deflected by small prisms. After passing through long pass filter (LPF), half-wave plate (HWP), interference filter (IF), photons are collected by coupling lens into single mode fibers (SMFs). The LPF from 650 nm is used to remove the scattering 404 nm pump light, and the IF with 3 nm bandwidth is used to ensure good spectral indistinguishability. The HWP and the fiber polarization controller (PC) can control the polarization state of photon in fiber to maintain V polarization. One way is inserted into a delay line to control the relative arrival time of two photons to ensure the temporal indistinguishability. The photons are injected into the chip through 4-channel input fiber array (FA) with the same 127 μm spacing and then collected from the chip to the output fiber array. After that, the output photons are detected by single photon counting modules (SPCMs) (Excelitas, SPCM-800-14-FC) and conveyed to the Time to Digital Converter (TDC) (ID800, IDQ) to conduct the coincidence counting of the output photon pairs.

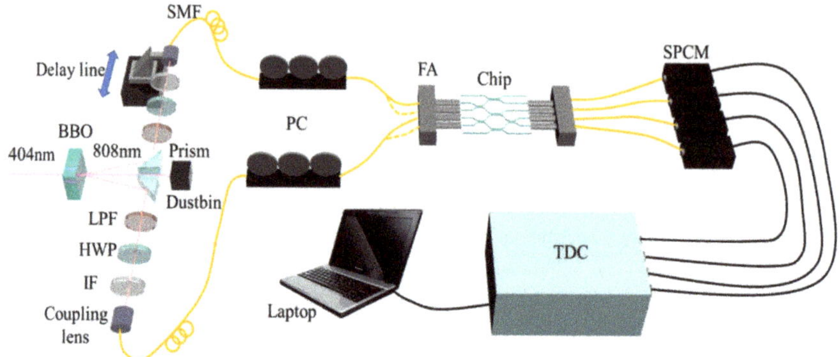

Figure 4. Experimental setup for quantum characterization. Through Type-I spontaneous parametric down-conversion (SPDC) process, 808 nm photon pairs are generated by pumping the BBO crystal using 140 mW, 404 nm CW diode laser. Long pass filter (LPF) from 650 nm and interference filter (IF) at 808 nm with 3 nm bandwidth are used to ensure spectral indistinguishability. Half-wave plate (HWP) and polarization controller (PC) are used to control the polarization state of photons in fiber. A delay line is inserted into one way to control the relative arrival time of photons to ensure temporal indistinguishability. Photons are injected into waveguides in the chip through fiber array and then collected at the output by another fiber array. Single photon counting modules (SPCMs) and the Time to Digital Converter (TDC) are used to conduct coincidence counting of different output-photon combinations.

By scanning the relative delay between two input photons, we can get coincidence counting curves for four kinds of output-photon combinations after post-selection. As shown in Figure 5a, when the input-photon combination is (1,3), the interference curve of output-photon combination (2′,4′) shows a HOM dip at interference point, where the relative delay is zero and the corresponding coincidence counts reduce to zero. Its HOM interference visibility is about 98.5 ± 1.2% with accidental coincidence counts subtracted. However, the interference curves of output-photon combinations (2′,3′) and (1′,4′) slightly change, and their coincidence counts at interference point get close to each other. In their interference curves, the occurrence of small dip or peak is due to the deviation of reflectivity for DCs and difference of internal phase between two arms of the MZ interferometer in the fabricated photonic circuit. The coincidence counts of the remaining output-photon combination (1′,3′) are close to zero. This represents that the post-selected output-photon state is a path-encoded entangled state:

$$|00\rangle_{ct} \Leftrightarrow |13\rangle_{ct} \rightarrow \frac{1}{\sqrt{2}}\left(|1'4'\rangle_{c't'} + e^{i\phi}|2'3'\rangle_{c't'}\right) \Leftrightarrow \frac{1}{\sqrt{2}}\left(|00\rangle_{c't'} + e^{i\phi}|11\rangle_{c't'}\right). \quad (2)$$

According to the theoretical prediction, the phase factor ϕ should be π, but we cannot determine its value directly in the chip. To completely analyze the output-photon state by acquiring the value of ϕ, we need to place more elements such as DCs and phase shifters behind the logic gates on the current chip [3], which need to lengthen the chip and fabricate the electrode or convert the encoding information of path to polarization outside the chip [35,44–46], which needs more optical elements and equipment. In Reference [1], the authors cannot determine the value of ϕ, but the demonstration of excellent logical basis operation of the CNOT gate and coherent quantum operation gives them great confidence. Similarly, as shown in Figure 5b–d, when the input-photon combinations are (1,4), (2,3), and (2,4), the visibilities of each interference curve for output-photon combinations (2′,3′), (2′,4′), and (2′,3′) are 97.8 ± 1.8%, 98.2 ± 1.8%, and 99.5 ± 0.5% with accidental coincidence counts subtracted, respectively. We can get three more path-encoded entangled state:

$$
\begin{aligned}
|01\rangle_{ct} &\Leftrightarrow |14\rangle_{ct} \rightarrow \frac{1}{\sqrt{2}}\left(|1'3'\rangle_{c't'} + e^{i\phi'}|2'4'\rangle_{c't'}\right) \Leftrightarrow \frac{1}{\sqrt{2}}\left(|01\rangle_{c't'} + e^{i\phi'}|10\rangle_{c't'}\right),\\
|10\rangle_{ct} &\Leftrightarrow |23\rangle_{ct} \rightarrow \frac{1}{\sqrt{2}}\left(|1'4'\rangle_{c't'} + e^{i\phi''}|2'3'\rangle_{c't'}\right) \Leftrightarrow \frac{1}{\sqrt{2}}\left(|00\rangle_{c't'} + e^{i\phi''}|11\rangle_{c't'}\right),\\
|11\rangle_{ct} &\Leftrightarrow |24\rangle_{ct} \rightarrow \frac{1}{\sqrt{2}}\left(|1'3'\rangle_{c't'} + e^{i\phi'''}|2'4'\rangle_{c't'}\right) \Leftrightarrow \frac{1}{\sqrt{2}}\left(|01\rangle_{c't'} + e^{i\phi'''}|10\rangle_{c't'}\right).
\end{aligned}
\quad (3)
$$

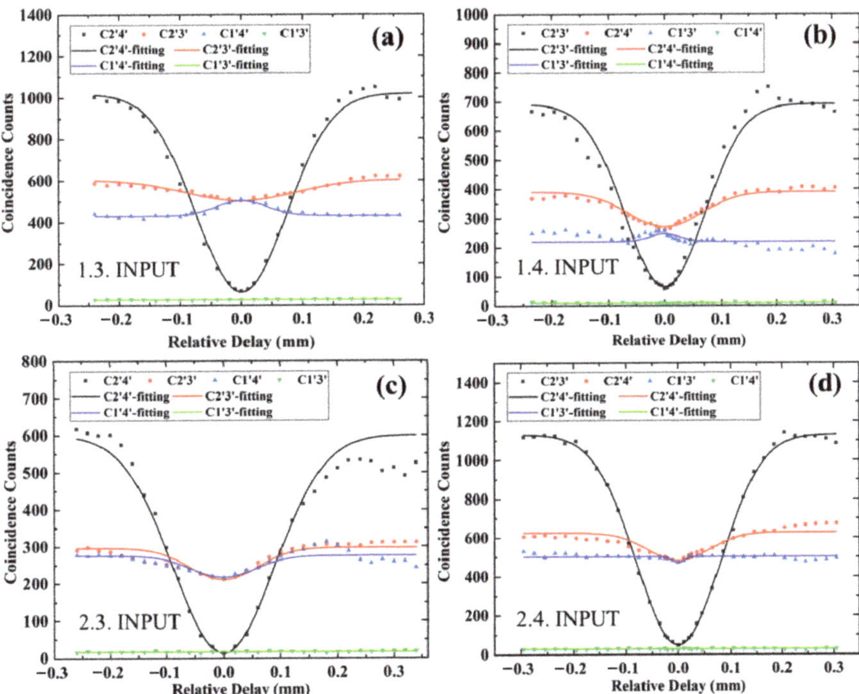

Figure 5. Coincidence counts in 30 s of post-selected output-photon combinations x′y′ denoted as Cx′y′ for different input-photon combinations xy denoted as x.y. INPUT as a function of the relative delay of photons input in x and y ports: (**a**) input (1,3); (**b**) input (1,4); (**c**) input (2,3); (**d**) input (2,4). From the interference curve with a deep dip, the HOM interference visibilities are (**a**) 98.5 ± 1.2%, (**b**) 97.8 ± 1.8%, (**c**) 98.2 ± 1.8%, (**d**) 99.5 ± 0.5%, respectively.

In theory, these phase factors should be $\phi' = 0$, $\phi'' = 0$ and $\phi''' = \pi$, but these phases cannot be confirmed directly either. In the future, we will make efforts to improve our experimental methods to conduct a complete characterization of the generated entangled states. In addition to the interference curves for different input-output combinations, we also need to reconstruct the truth table to determine the fidelity of this chip.

Compared with our previous work of path encoded CNOT gate [16], the success of this work depends not only on the high interference visibility but also on the equal probability distribution of two terms in the entangled state (Equations (2) and (3)), which requires tougher fabrication. For each input-photon combination, we normalize the coincidence counts for each output-photon combination with accidental counts subtracted to calculate their corresponding probability. As shown in Figure 6, the probabilities of each computational-basis output for each computational-basis input are represented by the height of the filled pink bars, and the height of empty bars stands for the theoretical value.

The reconstructed truth table of the chip coincides very well with the theoretical one and the average fidelity of the fabricated chip is as high as 98.8 ± 0.3%, which is also higher than 91.2 ± 0.2% in reference [1].

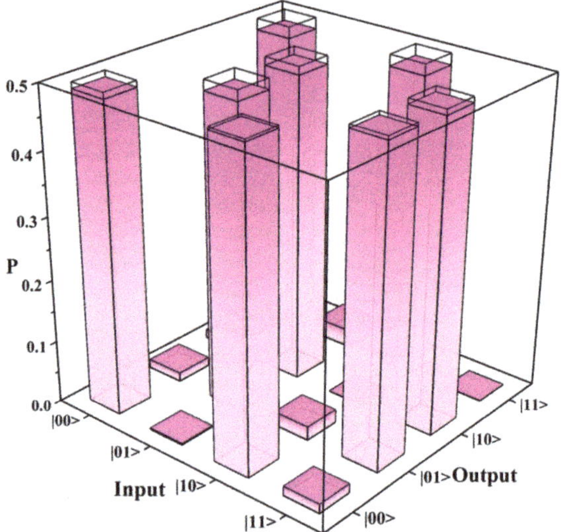

Figure 6. Reconstructed truth table of combined chip composed of one H gate and one CNOT gate. The labels on the Input axis represent $|C_q T_q\rangle$, and the labels on the Output axis represent $|C_q' T_q'\rangle$. P represents the probability for each input-output combination. The empty bars stand for the theoretical values, and the filled pink bars represent the experimental data. The average fidelity is as high as 98.8 ± 0.3%.

4. Discussion

The experimental results demonstrate that we can fabricate a photonic quantum chip constructed by cascading one H gate and one CNOT gate with high fidelity by FLDW technique. The H gate aims to prepare a superposition state of $|0\rangle$ and $|1\rangle$, and the post-selected CNOT gate generates a canonical two-qubit entangling gate. Therefore, this chip can generate all four path-encoded Bell entangled states. According to the classical characterization results, the performance of the fabricated chip can still be improved by more precisely controlled interaction lengths and a more stable micromachining system, but the improvement of fidelity is limited. The fabrication and measurement of this chip are more difficult than those of the CNOT gate, because the output probabilities of the two terms in the entangled state should be as equal as possible, which means that we need to make more efforts to control the fabrication details, adjust the alignment and coupling between fiber arrays and the chip, maintain the polarization of input photons and monitor the real-time detection process. Eventually, we successfully fabricated such a chip with a high fidelity of 98.8 ± 0.3%. This work is a demonstration of the cascading of one H gate and one CNOT gate. Furthermore, we can construct a quantum circuit to cascade and parallel several logic gates to generate more complex entangled states, such as GHZ state, but it requires more photons with quantum correlation [47,48], which is more difficult in measurement than in fabrication.

5. Conclusions

We realize the femtosecond laser direct writing of an integrated photonic quantum chip composed of one H gate and one CNOT gate for generating all four path-encoded Bell entangled states. For both classical and quantum characterization, the chip achieves great performance. The HOM interference visibilities of each input-output combination are all higher than 97.8 ± 1.8%, and the average fidelity of the reconstructed truth table is about 98.8 ± 0.3%, which suggest the high-quality of the fabricated chip. This basic quantum circuit is an important element and can be applied in many quantum computation algorithms, such as quantum Prime Factorization, quantum Grover Search and quantum Fourier Transform. Bell state is also an important entangled photon source, which is widely used in quantum communication and quantum computation. This work presents the capability of FLDW technique to fabricate more complex and functional photonic quantum computation chips.

Author Contributions: Conceptualization, Y.L., M.L., and Q.Z.; methodology, M.L. and Y.C.; software, Q.Z.; validation, M.L., Q.Z., and Y.C.; formal analysis, M.L.; investigation, M.L.; resources, Y.L., X.R., and Q.G.; data curation, M.L.; writing—original draft preparation, M.L.; writing—review and editing, Y.L., Y.C., and X.R.; visualization, M.L.; supervision, Y.L.; project administration, Y.L.; funding acquisition, Y.L. All authors have read and agreed to the published version of the manuscript.

Funding: This research was funded by The National Key R&D Program of China (2018YFB1107205, 2016YFA0301302); National Natural Science Foundation of China (NSFC) (61590933, 61590932, 11474010, 11527901); Joint Fund for Equipment Pre-research Space Science and Technology (6141B06140601); Strategic Priority Research Program of the Chinese Academy of Sciences (CAS) Chinese Academy of Sciences (XDB24030601) and the Fundamental Research Funds for the Central Universities.

Conflicts of Interest: The authors declare no conflict of interest.

References

1. Politi, A.; Cryan, M.J.; Rarity, J.G.; Yu, S.; Brien, J.L. Silica-on-Silicon Waveguide Quantum Circuits. *Science* **2008**, *320*, 646. [CrossRef] [PubMed]
2. Carolan, J.; Harrold, C.; Sparrow, C.; Martín-López, E.; Russell, N.J.; Silverstone, J.W.; Shadbolt, P.J.; Matsuda, N.; Oguma, M.; Itoh, M.; et al. Universal linear optics. *Science* **2015**, *349*, 711. [CrossRef] [PubMed]
3. Shadbolt, P.J.; Verde, M.R.; Peruzzo, A.; Politi, A.; Laing, A.; Lobino, M.; Matthews, J.C.F.; Thompson, M.G.; O'Brien, J.L. Generating, manipulating and measuring entanglement and mixture with a reconfigurable photonic circuit. *Nat. Photon.* **2012**, *6*, 45–49. [CrossRef]
4. Harris, N.C.; Bunandar, D.; Pant, M.; Steinbrecher, G.R.; Mower, J.; Prabhu, M.; Baehr-Jones, T.; Hochberg, M.; Englund, D. Large-scale quantum photonic circuits in silicon. *Nanophotonics* **2016**, *5*, 456–468. [CrossRef]
5. Taballione, C.; Wolterink, T.A.W.; Lugani, J.; Eckstein, A.; Bell, B.A.; Grootjans, R.; Visscher, I.; Geskus, D.; Roeloffzen, C.G.H.; Renema, J.J.; et al. 8x8 reconfigurable quantum photonic processor based on silicon nitride waveguides. *Opt. Express* **2019**, *27*, 26842–26857. [CrossRef] [PubMed]
6. Sahu, P.P.; Das, A.K. Polarization-Insensitive Thermo-Optic Mach Zehnder Device Based on Silicon Oxinitride Waveguide with Fast Response Time. *Fiber Integr. Opt.* **2009**, *29*, 10–20. [CrossRef]
7. Marshall, G.D.; Politi, A.; Matthews, J.C.; Dekker, P.; Ams, M.; Withford, M.J.; O'Brien, J.L. Laser written waveguide photonic quantum circuits. *Opt. Express* **2009**, *17*, 12546–12554. [CrossRef]
8. Meany, T.; Biggerstaff, D.N.; Broome, M.A.; Fedrizzi, A.; Delanty, M.; Steel, M.J.; Gilchrist, A.; Marshall, G.D.; White, A.G.; Withford, M.J. Engineering integrated photonics for heralded quantum gates. *Sci. Rep.* **2016**, *6*, 25126. [CrossRef]
9. Flamini, F.; Magrini, L.; Rab, A.S.; Spagnolo, N.; D'Ambrosio, V.; Mataloni, P.; Sciarrino, F.; Zandrini, T.; Crespi, A.; Ramponi, R.; et al. Thermally reconfigurable quantum photonic circuits at telecom wavelength by femtosecond laser micromachining. *Light Sci. Appl.* **2015**, *4*, e354. [CrossRef]
10. Fang, Z.W.; Haque, S.; Lin, J.T.; Wu, R.B.; Zhang, J.H.; Wang, M.; Zhou, J.X.; Rafa, M.; Lu, T.; Cheng, Y. Real-time electrical tuning of an optical spring on a monolithically integrated ultrahigh Q lithium nibote microresonator. *Opt. Lett.* **2019**, *44*, 1214–1217. [CrossRef]
11. Wu, R.B.; Lin, J.T.; Wang, M.; Fang, Z.W.; Chu, W.; Zhang, J.H.; Zhou, J.X.; Cheng, Y. Fabrication of a multifunctional photonic integrated chip on lithium niobate on insulator using femtosecond laser-assisted chemomechanical polish. *Opt. Lett.* **2019**, *44*, 4698–4701. [CrossRef] [PubMed]
12. Lin, J.T.; Bo, F.; Cheng, Y.; Xu, J.J. Advances in on-chip photonic devices based on lithium niobate on insulator. *Photon. Res.* **2020**, *8*, 1910–1936. [CrossRef]

13. Osellame, R.; Taccheo, S.; Marangoni, M.; Ramponi, R.; Laporta, P.; Polli, D.; De Silvestri, S.; Cerullo, G. Femtosecond writing of active optical waveguides with astigmatically shaped beams. *J. Opt. Soc. Am. B-Opt. Phys.* **2003**, *20*, 1559–1567. [CrossRef]
14. Cheng, Y.; Sugioka, K.; Midorikawa, K.; Masuda, M.; Toyoda, K.; Kawachi, M.; Shihoyama, K. Control of the cross-sectional shape of a hollow microchannel embedded in photostructurable glass by use of a femtosecond laser. *Opt. Lett.* **2003**, *28*, 55–57. [CrossRef] [PubMed]
15. Ams, M.; Marshall, G.D.; Spence, D.J.; Withford, M.J. Slit beam shaping method for femtosecond laser direct-write fabrication of symmetric waveguides in bulk glasses. *Opt. Express* **2005**, *13*, 5676–5681. [CrossRef] [PubMed]
16. Zhang, Q.; Li, M.; Chen, Y.; Ren, X.; Osellame, R.; Gong, Q.; Li, Y. Femtosecond laser direct writing of an integrated path-encoded CNOT quantum gate. *Opt. Mater. Express* **2019**, *9*, 2318–2326. [CrossRef]
17. Crespi, A.; Ramponi, R.; Osellame, R.; Sansoni, L.; Bongioanni, I.; Sciarrino, F.; Vallone, G.; Mataloni, P. Integrated photonic quantum gates for polarization qubits. *Nat. Commun.* **2011**, *2*, 566. [CrossRef]
18. Zeuner, J.; Sharma, A.N.; Tillmann, M.; Heilmann, R.; Gräfe, M.; Moqanaki, A.; Szameit, A.; Walther, P. Integrated-optics heralded controlled-NOT gate for polarization-encoded qubits. *NPJ Quantum Inf.* **2018**, *4*, 13. [CrossRef]
19. Crespi, A.; Osellame, R.; Ramponi, R.; Bentivegna, M.; Flamini, F.; Spagnolo, N.; Viggianiello, N.; Innocenti, L.; Mataloni, P.; Sciarrino, F. Suppression law of quantum states in a 3D photonic fast Fourier transform chip. *Nat. Commun.* **2016**, *7*, 10469. [CrossRef]
20. Spagnolo, N.; Aparo, L.; Vitelli, C.; Crespi, A.; Ramponi, R.; Osellame, R.; Mataloni, P.; Sciarrino, F. Quantum interferometry with three-dimensional geometry. *Sci. Rep.* **2012**, *2*, 862. [CrossRef]
21. Spagnolo, N.; Vitelli, C.; Aparo, L.; Mataloni, P.; Sciarrino, F.; Crespi, A.; Ramponi, R.; Osellame, R. Three-photon bosonic coalescence in an integrated tritter. *Nat. Commun.* **2013**, *4*, 1606. [CrossRef] [PubMed]
22. Ciampini, M.A.; Orieux, A.; Paesani, S.; Sciarrino, F.; Corrielli, G.; Crespi, A.; Ramponi, R.; Osellame, R.; Mataloni, P. Path-polarization hyperentangled and cluster states of photons on a chip. *Light Sci. Appl.* **2016**, *5*, e16064. [CrossRef] [PubMed]
23. Tang, H.; Di Franco, C.; Shi, Z.-Y.; He, T.-S.; Feng, Z.; Gao, J.; Sun, K.; Li, Z.-M.; Jiao, Z.-Q.; Wang, T.-Y.; et al. Experimental quantum fast hitting on hexagonal graphs. *Nat. Photon.* **2018**, *12*, 754–758. [CrossRef]
24. Xu, X.Y.; Huang, X.L.; Li, Z.M.; Gao, J.; Jiao, Z.Q.; Wang, Y.; Ren, R.J.; Zhang, H.P.; Jin, X.M. A scalable photonic computer solving the subset sum problem. *Sci. Adv.* **2020**, *6*, eaay5853. [CrossRef]
25. Sansoni, L.; Sciarrino, F.; Vallone, G.; Mataloni, P.; Crespi, A.; Ramponi, R.; Osellame, R. Two-Particle Bosonic-Fermionic Quantum Walk via Integrated Photonics. *Phys. Rev. Lett.* **2012**, *108*, 010502. [CrossRef]
26. Tang, H.; Lin, X.-F.; Feng, Z.; Chen, J.-Y.; Gao, J.; Sun, K.; Wang, C.-Y.; Lai, P.-C.; Xu, X.-Y.; Wang, Y.; et al. Experimental two-dimensional quantum walk on a photonic chip. *Sci. Adv.* **2018**, *4*, eaat3174. [CrossRef]
27. Crespi, A.; Osellame, R.; Ramponi, R.; Giovannetti, V.; Fazio, R.; Sansoni, L.; De Nicola, F.; Sciarrino, F.; Mataloni, P. Anderson localization of entangled photons in an integrated quantum walk. *Nat. Photon.* **2013**, *7*, 322–328. [CrossRef]
28. Rechtsman, M.C.; Zeuner, J.M.; Plotnik, Y.; Lumer, Y.; Podolsky, D.; Dreisow, F.; Nolte, S.; Segev, M.; Szameit, A. Photonic Floquet topological insulators. *Nature* **2013**, *496*, 196–200. [CrossRef]
29. Mukherjee, S.; Rechtsman, M.C. Observation of Floquet solitons in a topological bandgap. *Science* **2020**, *368*, 856. [CrossRef]
30. Wang, Y.; Lu, Y.-H.; Mei, F.; Gao, J.; Li, Z.-M.; Tang, H.; Zhu, S.-L.; Jia, S.; Jin, X.-M. Direct Observation of Topology from Single-Photon Dynamics. *Phys. Rev. Lett.* **2019**, *122*, 193903. [CrossRef]
31. Vest, G.; Rau, M.; Fuchs, L.; Corrielli, G.; Weier, H.; Nauerth, S.; Crespi, A.; Osellame, R.; Weinfurter, H. Design and Evaluation of a Handheld Quantum Key Distribution Sender module. *IEEE J. Sel. Top. Quant Electron.* **2015**, *21*, 131–137. [CrossRef]
32. Wang, C.-Y.; Gao, J.; Jiao, Z.-Q.; Qiao, L.-F.; Ren, R.-J.; Feng, Z.; Chen, Y.; Yan, Z.-Q.; Wang, Y.; Tang, H.; et al. Integrated measurement server for measurement-device-independent quantum key distribution network. *Opt. Express* **2019**, *27*, 5982–5989. [CrossRef] [PubMed]
33. Spagnolo, N.; Vitelli, C.; Bentivegna, M.; Brod, D.J.; Crespi, A.; Flamini, F.; Giacomini, S.; Milani, G.; Ramponi, R.; Mataloni, P.; et al. Experimental validation of photonic boson sampling. *Nat. Photon.* **2014**, *8*, 615–620. [CrossRef]

34. Crespi, A.; Osellame, R.; Ramponi, R.; Brod, D.J.; Galvão, E.F.; Spagnolo, N.; Vitelli, C.; Maiorino, E.; Mataloni, P.; Sciarrino, F. Integrated multimode interferometers with arbitrary designs for photonic boson sampling. *Nat. Photon.* **2013**, *7*, 545–549. [CrossRef]
35. Atzeni, S.; Rab, A.S.; Corrielli, G.; Polino, E.; Valeri, M.; Mataloni, P.; Spagnolo, N.; Crespi, A.; Sciarrino, F.; Osellame, R. Integrated sources of entangled photons at the telecom wavelength in femtosecond-laser-written circuits. *Optica* **2018**, *5*, 311–314. [CrossRef]
36. Lloyd, S. Almost Any Quantum Logic Gate is Universal. *Phys. Rev. Lett.* **1995**, *75*, 346–349. [CrossRef] [PubMed]
37. Deutsch, D.E.; Barenco, A.; Ekert, A. Universality in quantum computation. *Proc. R. Soc. Lond. Ser. A* **1995**, *449*, 669–677. [CrossRef]
38. Knill, E.; Laflamme, R.; Milburn, G.J. A scheme for efficient quantum computation with linear optics. *Nature* **2001**, *409*, 46–52. [CrossRef]
39. Okamoto, R.; O'Brien, J.L.; Hofmann, H.F.; Takeuchi, S. Realization of a Knill-Laflamme-Milburn controlled-NOT photonic quantum circuit combining effective optical nonlinearities. *Proc. Natl. Acad. Sci. USA* **2011**, *108*, 10067. [CrossRef]
40. Pan, J.-W.; Chen, Z.-B.; Lu, C.-Y.; Weinfurter, H.; Zeilinger, A.; Żukowski, M. Multiphoton entanglement and interferometry. *Rev. Mod. Phys.* **2012**, *84*, 777–838. [CrossRef]
41. Heilmann, R.; Gräfe, M.; Nolte, S.; Szameit, A. Arbitrary photonic wave plate operations on chip: Realizing Hadamard, Pauli-X, and rotation gates for polarisation qubits. *Sci. Rep.* **2014**, *4*, 4118. [CrossRef] [PubMed]
42. Politi, A.; Matthews, J.C.F.; Brien, J.L. Shor's Quantum Factoring Algorithm on a Photonic Chip. *Science* **2009**, *325*, 1221. [CrossRef] [PubMed]
43. Hong, C.K.; Ou, Z.Y.; Mandel, L. Measurement of subpicosecond time intervals between two photons by interference. *Phys. Rev. Lett.* **1987**, *59*, 2044–2046. [CrossRef] [PubMed]
44. Sansoni, L.; Sciarrino, F.; Vallone, G.; Mataloni, P.; Crespi, A.; Ramponi, R.; Osellame, R. Polarization Entangled State Measurement on a Chip. *Phys. Rev. Lett.* **2010**, *105*, 200503. [CrossRef]
45. Feng, L.-T.; Zhang, M.; Zhou, Z.-Y.; Chen, Y.; Li, M.; Dai, D.-X.; Ren, H.-L.; Guo, G.-P.; Guo, G.-C.; Tame, M.; et al. Generation of a frequency-degenerate four-photon entangled state using a silicon nanowire. *NPJ Quantum Inf.* **2019**, *5*, 90. [CrossRef]
46. Feng, L.-T.; Zhang, M.; Xiong, X.; Chen, Y.; Wu, H.; Li, M.; Guo, G.-P.; Guo, G.-C.; Dai, D.-X.; Ren, X.-F. On-chip transverse-mode entangled photon pair source. *NPJ Quantum Inf.* **2019**, *5*, 2. [CrossRef]
47. Zhang, M.; Feng, L.-T.; Zhou, Z.-Y.; Chen, Y.; Wu, H.; Li, M.; Gao, S.-M.; Guo, G.-P.; Guo, G.-C.; Dai, D.-X.; et al. Generation of multiphoton quantum states on silicon. *Light Sci. Appl.* **2019**, *8*, 41. [CrossRef]
48. Zhang, C.; Huang, Y.F.; Zhang, C.J.; Wang, J.; Liu, B.H.; Li, C.F.; Guo, G.C. Generation and applications of an ultrahigh-fidelity four-photon Greenberger-Horne-Zeilinger state. *Opt. Express* **2016**, *24*, 27059–27069. [CrossRef]

Article

AFM Analysis of Micron and Sub-Micron Sized Bridges Fabricated Using the Femtosecond Laser on YBCO Thin Films

Patrice Umenne 

Department of Electrical and Mining Engineering, University of South Africa, Florida 1709, South Africa; umennpo@unisa.ac.za; Tel.: +27-72-495-0922

Received: 10 November 2020; Accepted: 24 November 2020; Published: 8 December 2020 

Abstract: The research arose as a result of the need to use the femtosecond laser to fabricate sub-micron and nano-sized bridges that could be analyzed for the Josephson effect. The femtosecond laser has a low pulse duration of 130 femtoseconds. Hence in an optical setup it was assumed that it could prevent the thermal degradation of the superconductive material during fabrication. In this paper a series of micron and sub-micron sized bridges where fabricated on superconductive yttrium barium copper oxide (YBCO) thin film using the femtosecond laser, a spherical convex lens of focal length 30 mm and the G-code control programming language applied to a translation stage. The dimensions of the bridges fabricated where analyzed using the atomic force microscope (AFM). As a result, micron sized superconductive bridges of width 1.68 µm, 1.39 µm, 1.23 µm and sub-micron sized bridges of width 858 nm, 732 nm where fabricated. The length of this bridges ranged from 9.6 µm to 12.8 µm. The femtosecond laser technique and the spherical convex lens can be used to fabricate bridges in the sub-micron dimension.

Keywords: atomic force microscope; laser ablation diameter; separation distance (S_W); sub-micron bridges; YBCO thin film

---

1. Introduction

The main aim of this work was to fabricate sub-micron and nano sized bridges using the 775 nm wavelength femtosecond laser as it is used for ablative purposes in [1–4] on superconductive yttrium barium copper oxide (YBCO) thin film [5–8]. The resulting bridges could then be tested for the presence of the Josephson effect and used as Josephson junctions. In [9–13], several characteristics and applications of Josephson junctions are given in detail. The smaller the size of the bridges that are fabricated the more likely they are to show the Josephson effect. If proven to show the Josephson effect, they can then be used in applications such as superconducting quantum interference device (SQUID) as in [14] where they are used to detect the presence of magnetic moments. This papers focus is, however, restricted to several superconductive bridges fabricated in the achievable micron and sub-micron scale and their dimensional analysis using the atomic force microscope (AFM).

A series of bridges where fabricated using optical equipment such as reflective mirrors, iris diaphragm, spherical convex lens and optical techniques such as beam collimation and focusing to reduce the laser ablation diameter. Optimization techniques were used to reduce the laser ablation diameter by lowering the pulse energy of the laser [15] and to reduce the width of the bridge by reducing the distance between the laser ablation spots (S_w). In [15] several factors are discussed that can be used to control the minimum structure size of materials when machining with the femtosecond laser. G-code (RS-274) computer numerical control (CNC) programming language was used to enable the movement of the translation stage that holds the thin film sample. In the meantime, the laser was held in a stationary position while ablating the sample. In order to measure the dimensions of the bridges after fabrication the AFM [16–20] was utilized for imaging the sample. In [16–20], the AFM is utilized to scan and analyze, cells, molecules, superconductors, semiconductors and nanomaterials as examples to using the AFM. The AFM scans where done with type DT-NCHR diamond cantilever tips in tapping mode.

The main hypothesis in the research was the assumption that the low pulse duration (130 fs) of the femtosecond laser could reduce the thermal degradation of the superconductive YBCO thin film during ablation. This is required in the fabrication of Josephson junctions. However, a null hypothesis was achieved. Experiments showed that the pulse duration of the laser on the superconductive YBCO thin film during fabrication is actually a combined average of the feed rate of the translation stage which was set at 20 mm/min or 333 μms^{-1}, the frequency of the laser or pulse repetition rate in this case 1 kHz and the pulse duration of the laser which is 130 femtoseconds. As a result, the time spent by the laser on the YBCO sample during ablation is in fact much longer than femtoseconds. Hence some thermal degradation occurs. The main objective of the research was to use the femtosecond laser technique to fabricate micron, sub-micron and nano sized superconductive bridges on YBCO thin film that could be tested and used as Josephson junctions. The novelty of the research stems from the fact that the femtosecond laser has never been used previously to fabricate superconductive bridges on YBCO thin film that could be used as Josephson junctions. As a result, the effect of using the femtosecond laser for this purpose had not been previously analyzed.

2. Materials and Methods

The YBCO thin films utilized for the experiment were procured from ceraco ceramic GmbH company. The thin films had the following specifications: 9 by 9 mm YBCO film, single sided 200 nm thickness, one side polished. The YBCO thin films came on either an $LaAlO_3$ substrate or an MgO substrate. The thin films where the S-type smooth matrix useful for the manufacture of SQUIDS and were placed on a substrate of 10 by 10 by 0.5 mm dimension. The critical temperature of the YBCO thin film used was $T_C = 87$ K.

The laser beam was focused unto the YBCO sample to machine the bridges by using a spherical convex lens of 30 mm focal length. The femtosecond laser power ranged from (0–1000) mW. The laser was set at 2.1 mW in order to cut just slightly above the ablation threshold of the YBCO thin film and hence make the laser ablation diameter as small as possible. This optimization technique facilitates the fabrication of smaller bridges. The laser power setting of 2.1 mW in combination with the spherical convex lens of focal length 30 mm produced a laser ablation diameter of 15.8 µm.

Figure 1 shows the block diagram of the optical set up used to optimize the laser ablation spot size and to fabricate the bridges with the femtosecond laser. The set up consists of the laser source, the beam collimation set up, reflective mirrors, iris (manually adjustable aperture) and the spherical convex lens of 30 mm focal length.

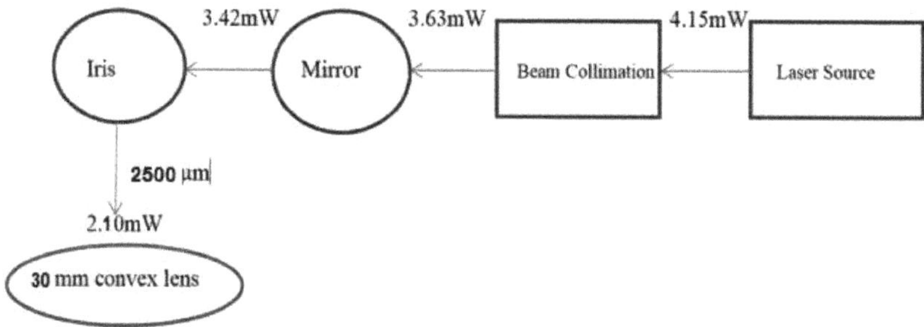

Figure 1. Block diagram of the overall laser optical experimental set up used to machine the bridges [21].

The function of the iris diaphragm is to remove unwanted sections of the laser in the outer periphery of the laser beam after the beam collimation process. The iris helps remove the poorly shaped sections of the laser beam produced by spherical aberration. This means only the central core of the laser beam would pass through to the focusing optics. The laser beam diameter from the laser source was 9.85 mm. The iris diaphragm could be adjusted from (0 –10,000) μm. The iris was set to a diameter of 2500 μm, thus reducing the laser beam diameter from 9.85 to 2.5 mm. The laser beam is then passed through the focusing optics and a laser ablation diameter of 15.8 μm is produced. During fabrication the YBCO sample is placed on a translation stage whose movement is programmed using G-code. The laser was kept stationary while the translation stage moved during the fabrication of the bridges.

The width of the bridges after fabrication is defined by the formula in Equation (1) [21]:

$$width\ of\ bridge = S_W - laser\ ablation\ diameter \tag{1}$$

where S_W is the distance between the laser ablation spots along the length of the square sample.

The length of the bridges is defined by the formula in Equation (2):

$$Length\ of\ bridge = |S_L - Laser\ ablation\ diameter| \tag{2}$$

where S_L is the distance between the laser ablation spots along the width of the square sample.

By bringing the laser ablation spots closer along the length of the sample the width of the bridge can be controlled. Similarly, by moving the laser ablation spots along the width of the sample you can control the length of the bridge fabricated. Initially the ablation strips are etched on the sample. The function of the ablation strips is to separate one bridge from another electronically. The ablation strip lines are fabricated by moving the laser vertically along the length of the YBCO square film. When the ablation strips are completed, they will have a width of 0.5 mm. The laser is moved vertically down the length of the YBCO thin film according to a specific factor. Horizontally into the sample, out of the sample and then vertically down by a factor. Across the width of the YBCO thin film, up along the length determined by a factor, again horizontally into the sample, back out and then back to the top along the length determined by a specific factor. This movement fabricates a bridge that is S-shaped and whose images are shown in the result section.

3. Results

Table 1 summarizes the bridges fabricated, the distance between the laser ablation diameters S_W set in the program, the laser ablation diameter and the width of the bridge that was machined. During the fabrication of all these bridges a conventional spherical convex lens of focal length 30 mm was used as the focusing optics.

Table 1. Summary of bridges fabricated and their dimensions.

Number	Name of Bridge	Separation Distance between Laser Ablation Spots (Sw)	Laser Ablation Diameter	Width of Bridge Formed
1	Micro-A	18 μm	15.8 μm	1.68 μm
2	Micro-B	17.5 μm	15.8 μm	1.39 μm
3	Micro-C	16.5 μm	15.8 μm	1.23 μm
4	SubMicro-D	16.5 μm	15.8 μm	858 nm
5	SubMicro-E	16 μm	15.8 μm	732 nm

3.1. AFM Analysis of Micro-A

When fabricating the bridge Micro-A, the distance between the laser ablation spots (S_W) along the length of the square sample was set at 18 μm. This setting determines the width of the bridge. The distance between the laser ablation spots (S_L) along the width of the square sample was set at approximately 5 μm. This setting determines the length of the bridge. The laser ablation diameter using the 30 mm focal length convex lens was 15.8 μm. As a result, a bridge of width 1.68 μm and length 12.79 μm was achieved as can be seen in the panel Figure 2. We focus on a very small scan area on the sample that is 20.5 by 20.5 μm.

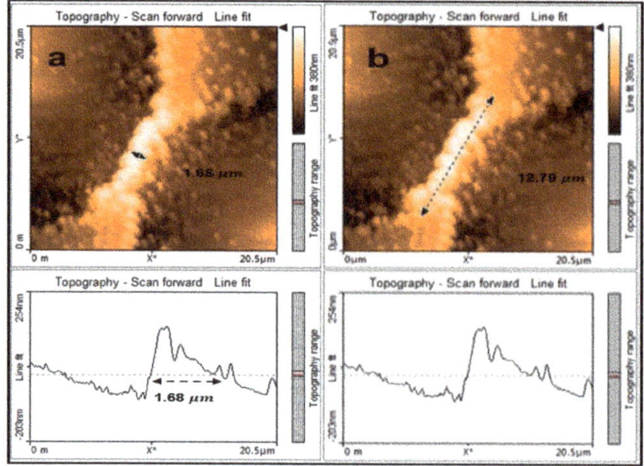

Figure 2. (**a**) topography line fit shows the width of the bridge at 1.68 μm and (**b**) topography line fit shows the length of the bridge at 12.79 μm.

The measurements are taken using the AFM as the cantilever tip scans transversely at 90 degrees to the laser ablation spot. By using the topography line fit shown in panel Figure 2a the width of the bridge is presented as an amplitude above the zero axis. The width of the amplitude shown in between the black arrows in the figure is 1.68 μm.

3.2. AFM Analysis of Micro-B

In the bridge Micro-B, the distance between the laser ablation spots (S_W) along the length of the square sample was set at 17.5 μm. The distance between the laser ablation spots (S_L) along the width of the sample was set at 5 μm, just as for Micro-A. The laser ablation diameter remained the same as 15.8 μm. A bridge of width 1.39 μm and length 12.26 μm was fabricated as can be seen in panel Figure 3. Again, the scan area was set at 20.5 by 20.5 μm. Therefore, by reducing the distance between the laser ablation spots (S_W) and keeping the laser ablation diameter constant one can produce a bridge of smaller width as per Equation (1).

The measurements were taken with an AFM. In the topography line fit in panel Figure 3a the width of the bridge is presented as an amplitude above the zero axis. The width of the amplitude in between the black arrows in the figure is 1.39 μm.

3.3. AFM Analysis of Micro-C

In the bridge Micro-C, the distance between the laser ablation spots (S_W) along the length of the square sample was set at 16.5 μm. The distance between the laser ablation spots (S_L) along the width of the sample was set at 5 μm, just as for Micro-A. The laser ablation diameter remained the same as 15.8 μm. Using the theory of Equation (1), a bridge of width 700 nm, is expected however, a bridge of width 1.23 μm is achieved as can be seen in Figure 4.

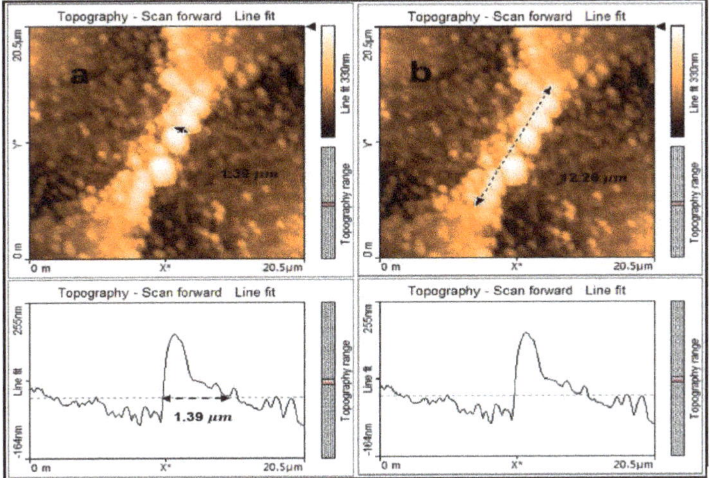

Figure 3. (**a**) topography line fit shows the width of the bridge at 1.39 μm and (**b**) topography line fit shows the length of the bridge at 12.26 μm.

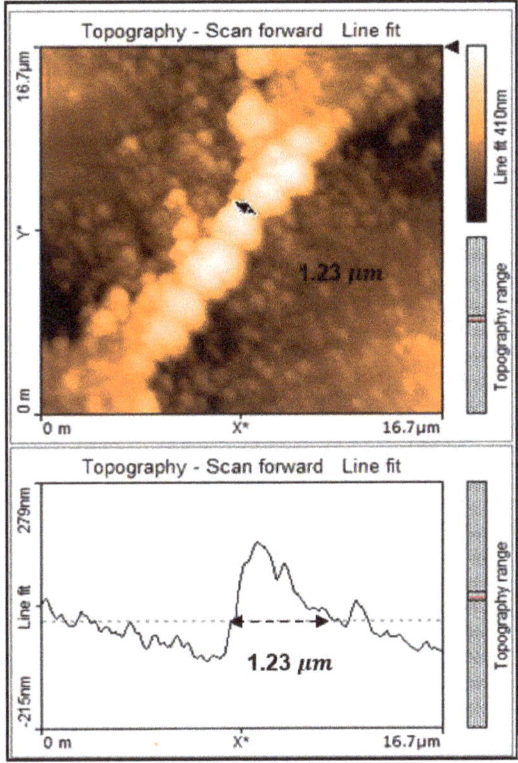

Figure 4. Topography line fit shows the width of the bridge at 1.23 μm.

In order to determine the exact width of this bridge the scan area was reduced to 16.7 by 16.7 μm.

3.4. AFM Analysis of SubMicro-D

For the bridge SubMicro-D, the distance between the laser ablation spots (S_W) along the length of the square sample was set at 16.5 µm. The distance between the laser ablation spots (S_L) along the width of the sample was set at 5 µm, just as for Micro-A. As a result, a bridge of width 858 nm and length 9.66 µm was fabricated as can be seen in the panel Figure 5.

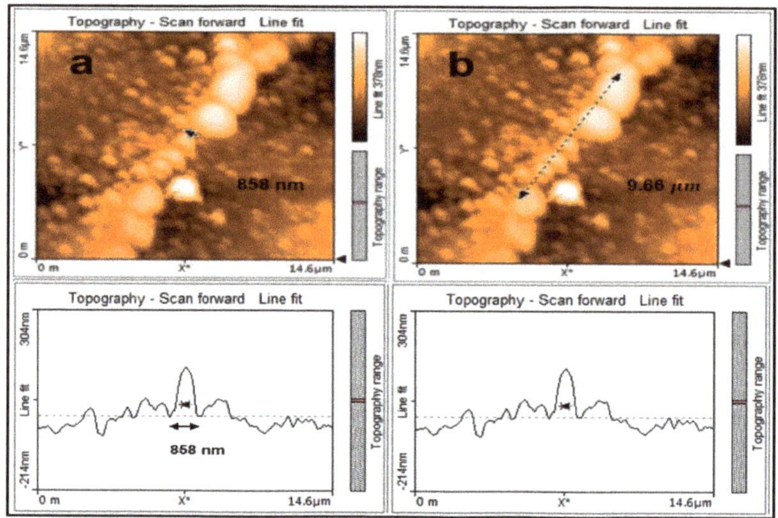

Figure 5. (**a**) topography line fit shows the width of the bridge at 858 nm and (**b**) topography line fit shows the length of the bridge at 9.66 µm.

In order to establish the dimensions of the bridge SubMicro-D the scan area was reduced to 14.6 by 14.6 µm.

3.5. AFM Analysis of SubMicro-E

In the design of the bridge SubMicro-E, the distance between the laser ablation spots (S_W) along the length of the square sample was set at 16 µm. The distance between the laser ablation spots (S_L) along the width of the sample was set at 5 µm, just as for Micro-A. A bridge of width 732 nm was achieved as can be seen in Figure 6. The scan area was set at 18.8 by 18.8 µm. The figure shows that if we machine any narrower the bridge can collapse. The reason is because we are approaching the diffraction limit of fabricating small structures with the femtosecond laser, which has a wavelength of 775 nm. In addition, in Figure 6 it can be seen that there is some thermal degradation on the YBCO thin film. This is shown by the different phases of ablation close to the bridge. There is a dark phase where there is more heating closer to the bridge and a lighter phase where there is less heating away from the bridge in the diameter of the laser ablation spot.

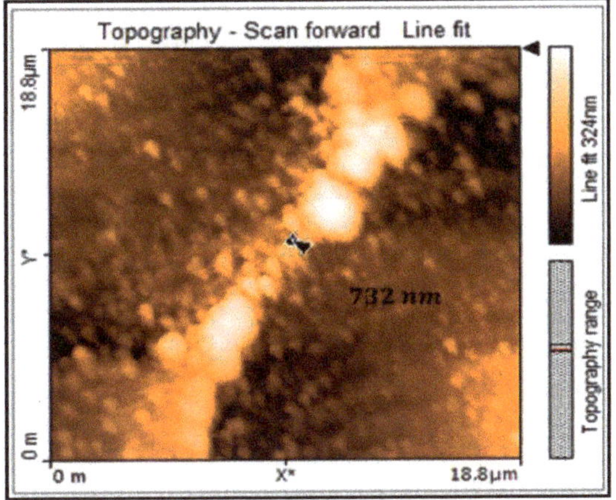

Figure 6. Topography line fit shows the width of the bridge at 732 nm.

4. Current-Voltage Characteristics (IVC's) of the Bridges

The I-V characteristic curves for the superconductive bridges Micro-A and Micro-B can be seen in Figure 7. Micro-A has a critical current I_C of 5.2 mA at temperature of 77 K and Micro-B has a critical current I_C of 4.2 mA at a temperature of 77 K. The critical current I_C for the superconductive bridge Micro-A slightly exceeds that of Micro-B. This can be explained by the fact that Micro-A is wider and longer than Micro-B, hence it is able to pass more current than Micro-B. Furthermore, from the AFM scans of Micro-A and Micro-B the geometrical boundaries of Micro-A are slightly better defined than that of Micro-B. This would mean that the superconductive phase of Micro-A is less damaged by the femtosecond laser during fabrication, hence produces more superconductive current at a temperature of 77 K. Critical currents I_C in the mA range are standard values of current at this temperature for this bridge dimensions.

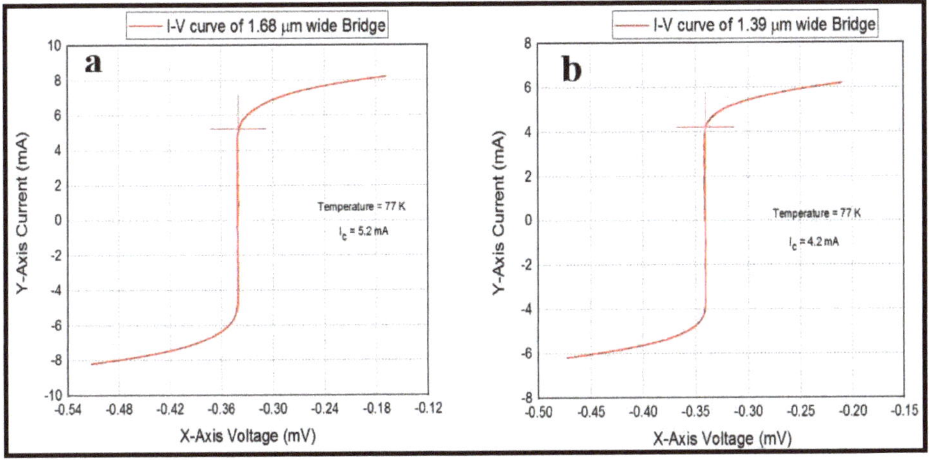

Figure 7. (**a**) I-V curve for Micro-A bridge and (**b**) I-V curve for Micro-B bridge.

5. Discussion

At the beginning of the research a hypothesis was made, that by using the femtosecond laser to machine the superconductive bridges thermal degradation of the superconductive sample could be reduced, since the femtosecond laser has a low pulse duration of 130 femtoseconds. However, a null hypothesis was achieved because the average time the laser spends on the sample depends not only on the pulse duration of the laser (130 fs), but also on the frequency of the laser (1 kHz) and the translation stage feed-rate (20 mm/min). Thermal degradation of the sample is represented by the dark phases in Figure 6 close to the bridge where there is more heating from the laser. Further away from the bridge there is a light phase indicative of less heating from laser and less thermal damage. When the YBCO thin film is less damaged by the heat from the laser it maintains a light phase seen on the AFM scans. The effect of the thermal degradation on the YBCO thin film is that the material changes from a superconductive phase to a non-superconductive phase that is either a resistor or an electronically open material. In the non-superconductive phase, it is impossible to produce a Josephson junction which was the original objective of the research. All the bridges fabricated in this research have an intact superconductive phase and could possibly be used as Josephson junctions. This is indicated in the I-V curve characteristics. The femtosecond laser technique was used instead of micro-milling techniques since micro-milling techniques require contact and use friction which produces heat and requires work to be done at temperatures that do not exceed 35 °C. Such temperatures would easily damage the superconductive YBCO material which is sensitive to high temperatures and can operate at a temperature that does not exceed 87 K (−186 °C). Moreover, the superconductive YBCO material requires non-contact techniques during fabrication while micro-milling is a contact method. Finally, the micro-milling technique generally can fabricate only micron sized dimensions while sub-micron and nano-scales are necessary to see the Josephson effect. The femtosecond laser technique is non-contact, can fabricate very small structures and generally has a very low-pulse duration on average that can evade heating of the sample.

6. Conclusions

In conclusion three micron-sized bridges; Micron-A, Micron-B, Micron-C and two sub-micron sized bridges; SubMicron-D and SubMicron-E were fabricated on superconductive YBCO thin film for possible use as a Josephson junction. These bridges were analyzed for their dimensions with the aid of the AFM. In the end it was discovered that the smaller the distance between the laser ablation spots S_W along the length of the square sample, the smaller the width of the bridge that is fabricated. This is the case if the laser ablation diameter is kept constant. There is however a limitation on the smallest width of the bridge that can be fabricated which will depend on the wavelength of the laser. Furthermore, a certain amount of thermal degradation occurs on the YBCO bridge even when using the femtosecond laser with a low pulse duration.

Funding: This research was funded by the University of South Africa. The APC was funded by the University of South Africa and the ESKOM South Africa TESP Grant.

Conflicts of Interest: The authors declare no conflict of interest.

References

1. Vogel, A.; Noack, J.; Huttman, G.; Paltauf, G. Mechanisms of femtosecond laser nanosurgery of cells and tissues. *Appl. Phys. B* **2005**, *81*, 1015–1047. [CrossRef]
2. Yao, J.; Zhang, C.; Liu, H.; Dai, Q.; Wu, L.; Lan, S.; Gopal, A.V.; Trofimov, V.A.; Lysak, T.M. Selective appearance of several laser-induced periodic surface structure patterns on a metal surface using structural colors produced by femtosecond laser pulses. *Appl. Surf. Sci.* **2012**, *258*, 7625–7632. [CrossRef]
3. Wang, S.Y.; Ren, Y.; Cheng, C.W.; Chen, J.K.; Tzou, D.Y. Micromachining of copper by femtosecond laser pulses. *Appl. Surf. Sci.* **2013**, *265*, 302–308. [CrossRef]

4. Bian, Q.; Yu, X.; Zhao, B.; Chang, Z.; Lei, S. Femtosecond laser ablation of indium tin-oxide narrow grooves for thin film solar cells. *Opt. Laser Technol.* **2013**, *45*, 395–401. [CrossRef]
5. Campbell, T.A.; Haugan, T.J.; Maartense, I.; Murphy, J.; Brunke, L.; Barnes, P.N. Flux pinning effects of Y_2O_3 nanoparticulate dispersions in multilayered YBCO thin films. *Phys. C* **2005**, *423*, 1–8. [CrossRef]
6. Wang, W.T.; Li, G.; Pu, M.H.; Sun, R.P.; Zhou, H.M.; Zhang, Y.; Zhang, H.; Yang, Y.; Cheng, C.H.; Zhao, Y. Chemical solution deposition of YBCO thin film by different polymer additives. *Physica C* **2008**, *468*, 1563–1566. [CrossRef]
7. Guo, L.S.; Chen, Y.Y.; Cheng, L.; Li, W.; Xiong, J.; Tao, B.W.; Yao, X. Liquid phase epitaxy of REBCO (RE=Y, Sm) thick films on YBCO thin film deposited on LAO substrate. *J. Cryst. Growth* **2013**, *366*, 47–50. [CrossRef]
8. Jacob, M.V.; Mazierska, J.; Savvides, N.; Ohshima, S.; Oikawa, S. Comparison of microwave properties of YBCO films on MgO and $LaAlO_3$. *Physica C* **2002**, *372–376*, 474–477. [CrossRef]
9. Ryu, C.; Blackburn, P.W.; Blinova, A.A.; Boshier, M.G. Experimental Realization of Josephson Junctions for an atom SQUID. *Phys. Rev. Lett.* **2013**, *111*, 205301. [CrossRef] [PubMed]
10. Jeanneret, B.; Benz, S.P. Application of the Josephson effect in electrical metrology. *Eur. Phys. J. Spec. Top.* **2009**, *172*, 181–206. [CrossRef]
11. Ren, H.; Pientka, F.; Hart, S.; Pierce, A.T.; Kosowsky, M.; Lunczer, L.; Schlereth, R.; Scharf, B.; Hankiewicz, E.M.; Molenkamp, L.W.; et al. Topological superconductivity in a phase-controlled Josephson junction. *Nature* **2019**, *569*, 93–98. [CrossRef] [PubMed]
12. Cassidy, M.C.; Bruno, A.; Rubbert, S.; Irfan, M.; Kammhuber, J.; Schouten, R.N.; Akhmerov, A.R.; Kouwenhoven, L.P. Demonstration of an ac Josephson junction laser. *Science* **2017**, *355*, 939–942. [CrossRef] [PubMed]
13. Dana, S.K.; Sengupta, D.C.; Edoh, K.D. Chaotic dynamics in josephson junction. *IEEE Trans. Circuits Syst. I Fundam. Theory Appl.* **2001**, *48*, 990–996. [CrossRef]
14. Bouchiat, V. Detection of magnetic moments using a nano-SQUID: Limits of resolution and sensitivity in near-field SQUID magnetometry. *Supercond. Sci. Technol.* **2009**, *22*, 064002. Available online: https://iopscience.iop.org/article/10.1088/0953-2048/22/6/064002 (accessed on 14 May 2009). [CrossRef]
15. Korte, F.; Serbin, J.; Koch, J.; Egbert, A.; Fallnich, C.; Ostendorf, A.; Chichkov, B.N. Towards nanostructuring with femtosecond laser pulses. *Appl. Phys. A* **2003**, *77*, 229–235. [CrossRef]
16. Gross, L.; Mohn, F.; Moll, N.; Liljeroth, P.; Meyer, G. The chemical structure of a molecule resolved by atomic force microscopy. *Science* **2009**, *325*, 1110–1114. [CrossRef] [PubMed]
17. Muller, D.J.; Dufrene, Y.F. Atomic force microscopy: A nanoscopic window on the cell surface. *Trends Cell Biol.* **2011**, *21*, 461–469. [CrossRef] [PubMed]
18. Li, N.; Li, Z.; Ding, H.; Ji, S.; Chen, X.; Xue, Q.K. An atomic force microscopy study of single-layer FeSe superconductor. *Appl. Phys. Express* **2013**, *6*, 113101. [CrossRef]
19. Gonnelli, R.S. Atomic force microscopy in the surface characterization of semiconductors and superconductors. *Philos. Mag. B* **2000**, *80*, 599–609. [CrossRef]
20. Koblischka, M.R.; Winter, M.; Hartmann, U. Nanostripe structures in $SmBa_2Cu_3O_x$ superconductors. *Supercond. Sci. Technol.* **2007**, *20*, 681–686. [CrossRef]
21. Umenne, P. Fabrication of Nano Josephson Junctions Using the Femtosecond Laser Technique on High T_C Superconducting $YBa_2Cu_3O_7$ thin Films. Ph.D. Thesis, University of South Africa, Pretoria, South Africa, February 2018. Available online: http://hdl.handle.net/10500/23646 (accessed on 27 February 2018).

Publisher's Note: MDPI stays neutral with regard to jurisdictional claims in published maps and institutional affiliations.

Article

Fabrication of a 3D Multi-Depth Reservoir Micromodel in Borosilicate Glass Using Femtosecond Laser Material Processing

Ebenezer Owusu-Ansah * and Colin Dalton

Department of Electrical & Computer Engineering, Schulich School of Engineering, University of Calgary, 2500 University Drive NW, Calgary, AB T2N 1N4, Canada; cdalton@ucalgary.ca

* Correspondence: eowusuan@ucalgary.ca

Received: 13 November 2020; Accepted: 5 December 2020; Published: 6 December 2020

Abstract: Micromodels are ideal candidates for microfluidic transport investigations, and they have been used for many applications, including oil recovery and carbon dioxide storage. Conventional fabrication methods (e.g., photolithography and chemical etching) are beset with many issues, such as multiple wet processing steps and isotropic etching profiles, making them unsuitable to fabricate complex, multi-depth features. Here, we report a simpler approach, femtosecond laser material processing (FLMP), to fabricate a 3D reservoir micromodel featuring 4 different depths—35, 70, 140, and 280 μm, over a large surface area (20 mm × 15 mm) in a borosilicate glass substrate. The dependence of etch depth on major processing parameters of FLMP, i.e., average laser fluence (LF_{av}), and computer numerically controlled (CNC) processing speed (PS_{CNC}), was studied. A linear etch depth dependence on LF_{av} was determined while a three-phase exponential decay dependence was obtained for PS_{CNC}. The accuracy of the method was investigated by using the etch depth dependence on PS_{CNC} relation as a model to predict input parameters required to machine the micromodel. This study shows the capability and robustness of FLMP to machine 3D multi-depth features that will be essential for the development, control, and fabrication of complex microfluidic geometries.

Keywords: micromodels; porous media; 3D multi-depth channels; laser machining; femtosecond laser micromachining; femtosecond laser material processing; micro/nanotechnology fabrication

---

1. Introduction

The use of micromodels, also known as porous media, for microfluidic transport investigations has been extensively studied in the literature for many applications, such as oil recovery [1–5] and carbon dioxide storage [6–11] processes. For example, silicon and glass-based micromodels have been used to study pore-scales to understand oil-water-solid interactions, multiphase flow, and the dynamics of microemulsions in enhanced oil recovery processes [4,5,11,12]. This is due to the ability to fabricate micromodels to mimic the three dimensional, multiple depths naturally occurring in porous media (such as oil-bearing rock formations), and the ease to integrate them with optical instruments for real time and in-situ observation of complex flow behaviour [13]. Naturally occurring porous media consist of complex 3D (multiple depth) networks of pores and throats that makes them challenging to study with 2D (uniform depth) micromodels, as the physics of the third dimension, which are critical for understanding flow in porous media, cannot be captured. For example, oil and bubble break-up in multiphase flow is largely dependent on capillary snap-off, a mechanism known to occur when sizes of throats are smaller than pore bodies in the two dimensions that are perpendicular to the flow direction, making it difficult for multiphase flow investigations using 2D micromodels [14–17]. Therefore, 3D micromodels are essential for studying transport in porous media, including emulsion flow, two phase displacement, three-phase flow, foam flow, etc., that has high dependence on capillary effects.

The most widely used non-additive manufacturing and conventional method to fabricate 3D micromodels is photolithography, which involves the transfer of a predesigned pattern from a mask to a substrate material, typically glass, followed by a wet chemical etch to define the features in the glass [17]. The process involves several wet processing steps, the need for photomasks, and complicated multi step processes requiring many items of fabrication equipment. Also, the approach suffers from mask undercut due to isotropic etching of substrate by the etchant (typically hydrofluoric acid, HF, for glass) that negatively impacts the ability to control etch feature sizes and resolution, making it difficult to fabricate 3D features with multiple depths in the same substrate [17,18]. Recent progress in wet photolithography includes the work of Xu et al., who fabricated a two-depth 3D micromodel in the same glass substrate by varying the depth difference between the pore body and throat [16]. Also, Yun et al. used a similar approach to achieve two depths in silicon by repeating the etch process twice [19]. These are time consuming processes, requiring multiple masks, and provide little to no control on lateral separation between etched features. The fabrication of micromodels using dry etching photolithographic methods, where the photoresist or masking material is exposed to a plasma of reactive gasses such as Cl_2, O_2, and BCl_3, to remove the unprotected substrate material, has been reported [17,20–22]. In comparison to wet etching, dry etching methods, such as reactive ion etching (RIE), allows for control on the etch direction that results in vertical channel sidewalls; however, RIE requires sophisticated facilities [17] and is also limited to the fabrication of 2D channels (i.e., a single uniform depth throughout) [21,22]. On the other hand, additive manufacturing methods, e.g., stereolithography or 3D printing, can be used to make 3D micromodels from many materials, including resins, polymers, and hydrogels; however, they are limited to larger than micron sized features due to the spatial requirements for solidification of the liquid materials and are typically not optically transparent [17,23–25].

Femtosecond (fs) laser material processing (FLMP) is a simpler approach that has been used to date to machine 2D microchannel features into optically transparent materials such as borosilicate glass [26]. Others have used FLMP together with wet etch processes to produce 2D structures in photosensitive glass substrates [27]. Here in this study, we showed the capability of using FLMP with no additional wet etch methods to fabricate 3D microstructures consisting of 4 different depths in the same borosilicate glass substrate for use as a reservoir micromodel. Details of the FLMP method and its advantages over widely used conventional micro/nanotechnology (MNT) fabrication approaches are given in the next section.

Femtosecond Laser Material Processing (FLMP)

FLMP technique allows the development, control, and fabrication of MNT systems such as microfluidic and lab-on-a-chip devices that are not easily accomplished with traditional methods, such as photolithography [28–30]. Unlike conventional MNT fabrication, in FLMP, there is no need for photomasks or multiple coating and chemical etching procedures [28]. FLMP involves a computer numerically controlled (CNC) motion that can machine complex patterns through cycles of focused laser beam passes with high precision. Applications of FLMP includes the fabrication of microfluidic devices such as micro- [31,32], hydro-dynamic fluid pumps [29,33], and dielectrophoretic assays [30,34]. Also, FLMP has allowed internal machining of quartz to create waveguides [35–37] in optical systems, and the fabrication of complex X-ray masks in thin sheets of tungsten, a material that is not suited for chemical based etch methods due to its non-uniform structure that leads to uneven etch profiles [38].

When a femtosecond (fs) laser pulse incidents on a material, photon absorption occurs on a timescale ($\sim 10^{-14}$–10^{-13} s) that is shorter than the electron-phonon coupling relaxation process ($\sim 10^{-12}$–10^{-11} s), delivering energy to the electrons while leaving the ions and the lattice "cold". This ensures that within the duration of the fs pulse, there is little to no thermal energy transfer to the lattice which decouples the optical absorption processes from lattice thermalization processes.

The energy absorbed by the electrons causes excitation which breaks the bonds formed by these electrons with minimal heating of the material substrate [39]. In comparison to long laser pulses (e.g., nanosecond) [40–42], fs laser pulses produce high peak electric fields (~10^{12} V/m) which are approximately 3 orders of magnitude greater than the electric field (10^9 V/m) that binds electrons to atoms [39,43]. This make fs laser processing versatile to process a wide variety of materials including optically opaque and transparent materials, such as metals, glass, and silicon wafers. The high peak electric field allows non-linear optical absorption processes, such as multiphoton absorption and tunneling ionization [44], within the material substrate when the laser beam is tightly focused, resulting in bond breakage and the ablation of material from the exposed surface. The wavelength of most fs lasers used for FLMP of wide bandgap materials (e.g., semiconductors and glass) is typically $>$ 750 nm (1.65 eV), making the process nearly wavelength independent as the bandgap of these materials are mostly higher (~>2 eV) than the photon energy [45]. When the energy of an incident photon is larger than the bandgap of the material substrate, absorption occurs, and electrons are excited to the conduction band. On the contrary, optical absorption by electrons does not occur when the photon energy is smaller than the bandgap. However, when light with large peak electric fields, such as those generated by fs laser beams, are focused to produce an extremely high density of photons, electron absorption is possible through multiple photon absorption at several virtual states. This multiphoton absorption process allows electron excitation into the conduction band. For ablation to occur, the density of free electrons in the conduction band should reach a critical density that is achieved beyond a threshold laser fluence (optical breakdown) which is material dependent [46].

The FLMP technique is ideally suited to micro-structuring, as the ultra-short pulse width of the fs laser is shorter than the thermal diffusion times of most materials, including metals, ceramics, and glass [47–49]. In FLMP, the formation of a heat affected zone (HAZ), when a large portion of a laser pulse's energy is transformed to heat around the irradiated area and causes material damage, is significantly suppressed. Thus, with FLMP, there is little or no HAZ around the exposed site, resulting in less damage to the substrate material than conventional CO_2, nanosecond, and long pulse lasers. This allows for fine control of feature sizes not possible with CO_2 and long pulse lasers, enabling the fabrication of high-precision and high-quality MNT devices [28,39,44,50,51]. Also, an additional feature of the FLMP technique is the ability to easily make changes to a design by modifying the machining pattern on a computer, i.e., editing a CAD file. This significantly reduces the cost of prototyping by removing the need for multiple high-resolution photomasks and allows for a fast-iterative design cycle.

Recent investigations in FLMP have included efforts on how to effectively control the processing parameters, such as CNC speed, fluence (energy density), focused laser beam size, wavelength, and repetition rate [47,52–57]. These processing parameters have a significant effect on the properties of the resultant material etch parameters, such as etch profile (including cleanliness of the cut-edge), depth, feature size resolution, and surface roughness. Kam et al. used FLMP to machine multi-depth microchannel networks onto a silicon substrate for use as a gas exchanger [52]. It was found that the processing speed had a significant effect on the surface quality and the processing time. Hayden studied a simple 3D computer simulation tool to help predict some of the resultant etch parameters of FLMP on sodalime glass, borosilicate glass, and silicon substrates [47]. These investigations are important to harness any latent potential of the FLMP technique. Here, a study on the effect of average laser fluence (LF_{av}), and CNC processing speed (PS_{CNC}) to determine their relationship with the resultant etch depth in a borosilicate glass substrate is presented. The obtained relations were then used as models to guide the fabrication of 3D multi-depth features into a borosilicate glass substrate with 4 different depths for use as a reservoir micromodel.

2. Materials and Methods

A detailed description of the FLMP workstation setup used for this study has previously been reported [28,47]. However, for convenience and minor changes in the optical path, a brief description is given here. A schematic representation of the FLMP workstation is shown in Figure 1. It consists of a Ti:Sapphire Regenerative Amplifier Laser System (Spectra-Physics, Spitfire Pro, USA) that produces 800 nm infra-red (IR) radiation with 100 femtosecond (fs) pulse duration. The maximum output laser power arriving at the working piece substrate was 2.5 W when measured with a power meter (Ophir Meter) at a repetition rate of 1 kHz. The wavelength was tunable from 780–820 nm, while the repetition rate could be varied from 0.1–1 kHz. A summary of the input processing parameters used in this study are given in Table 1. The laser system was synchronized to a CNC stage (Aerotech, Inc., Pittsburgh, PA, USA) that allows XYZθ motions. Precise motion control, positioning, and machining were possible over a large area of 150 mm × 150 mm.

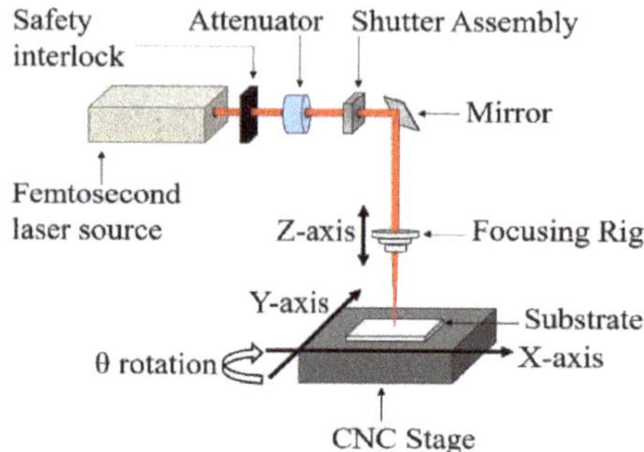

Figure 1. Schematic illustration of the femtosecond laser material processing (FLMP) setup. The path of the laser beam is fixed while the CNC stage allows XYZθ motions.

Table 1. Processing parameters used for FLMP of borosilicate glass substrate.

Processing Parameters	Value	Unit
Pulse width (τ)	100	fs
Wavelength (λ)	800	nm
Repetition rate	1	kHz
Beam diameter	12.3	μm
Pitch (center-to-center) of beam	5	μm
CNC speed (PS_{CNC})	0.025–10	mm/s
Average laser fluence (LF_{av})	14.31–1388.62	J/cm^2
Resultant etch depth	3.5–223.8	μm

The pattern to be machined was first designed using CAD/CAM software (Alphacam 2019 R1) and converted to a G-code text file, which was uploaded onto a computer that controls the motion stages. The path of the CNC motion stage enables the laser beam, when on, to create the desired pattern in the substrate located on the workpiece. This allows programmable, accurate, and repeatable motions for patterning complex MNT features. The laser beam path, which is fixed, was directed through a set of optical components, including safety interlock, attenuator, opto-mechanical shutter assembly, mirrors, and a focusing lens (housed in a focus rig) onto the CNC motion stage. The vertical Z axis motion allows the laser beam to be focused on different thicknesses of material substrates with the aid of an alignment camera and light mounted above the focus rig. The borosilicate glass substrate was held in place on the CNC stage by a vacuum suction source. Material properties of the borosilicate glass (McMaster-Carr®, Part # B84760365) were; density: 2440 kg/m^3, hardness: Knoop 418 KHN100, refractive index: 1.47, and the two largest components by % composition were SO_2: 70–87% and BO_3: 1–20%.

The attenuator was used to control the amount of laser energy arriving at the material substrate. During FLMP, the laser beam was always on, and therefore, the shutter assembly was needed to block off the beam when no machining was required, especially when the CNC stage was moving to a new location to machine a new feature on the substrate. The overhead camera and light were used for alignment purposes. An exhaust was mounted near the laser beam-substrate surface to remove machined debris during all FLMP experiments. To investigate etch depth dependence on LF_{av}, and PS_{CNC}, square features (1500 μm × 1500 μm) were machined in borosilicate glass substrates where all FLMP parameters were held constant while varying LF_{av} and PS_{CNC}, respectively. To machine features larger than the focused laser beam size, toolpaths consisting of several lines were generated for each feature. The pitch, spacing between the toolpath lines (center-to-center), was experimentally determined as it affects the machining time and the roughness of the etched surface. The extensive data on the effect of pitch variations on material substrate roughness will be covered in another manuscript. A 5 μm pitch and single laser beam pass were used for this work unless stated otherwise. After laser machining, the borosilicate glass substrates were immersed in an isopropyl alcohol sonication bath for 30 mins to remove remaining debris before a contact surface profilometer (P-6, KLA Tencor) with a 2 μm tip was used to characterize the etch profiles. Optical microscopy images were taken with Mitutoyo (Ultraplan FS110) while regression analysis data fitting was performed using OriginPro® software (version 8).

3. Results and Discussion

3.1. FLMP Etch Profiles

The size of the focused laser beam spot was experimentally determined by systematically varying the vertical Z position of the focus lens (f = 25 mm, F/0.6, Edmund Optics®) to machine 4 mm-length line features on the borosilicate glass substrate. At PS_{CNC} of 0.25 mm/s and 0.617 mJ power, an expected Gaussian-like etch profile was produced, as shown in the 2D line profile scan in Figure 2. For Gaussian-like profiles, the full-width at half-maximum (FWHM) value of 12.3 μm was determined as the diameter of the focused laser beam [58]. This was used to calculate the circular area in the determination of all LF_{av} values reported in this work.

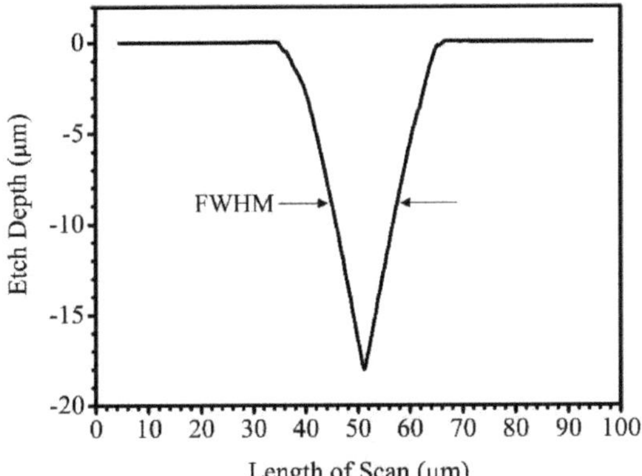

Figure 2. Line profile scan across 4 mm-length line feature machined into borosilicate glass using FLMP at PS_{CNC} of 0.25 mm/s and 0.617 mJ power. The profile was recorded at 2 μm/s, 10 Hz, and 2 mg applied force. The full-width at half-maximum (FWHM) value of 12.3 μm of the Gaussian-like etched profile was determined as the spot size of the focused laser beam diameter.

The profile of an etched area covering 1500 μm × 1500 μm was also etched at a pitch of 5 μm, PS_{CNC} of 0.1 mm/s, and LF_{av} of 329.06 J/cm^2. Figure 3 is a line profile scan across the etched area. The profile shows two inclined lines that reveal a symmetrical (isosceles) trapezoid geometry in comparison to the vertical lines of a rectangle. This was a direct consequence of the Gaussian-like profile of the focused laser beam as shown in Figure 2. The difference between the programmed G-code width of 1500 μm and the resultant machined width of ~1540 μm represents an offset value of ~40 μm that could be accounted for during subsequent CAD/CAM designs. However, no offset in width was factored into the designs reported in this work since that was not the focus of the study. The analysis of the trapezoid geometry also showed that the inclined etch surfaces make ~8° contact angle with the vertical plane.

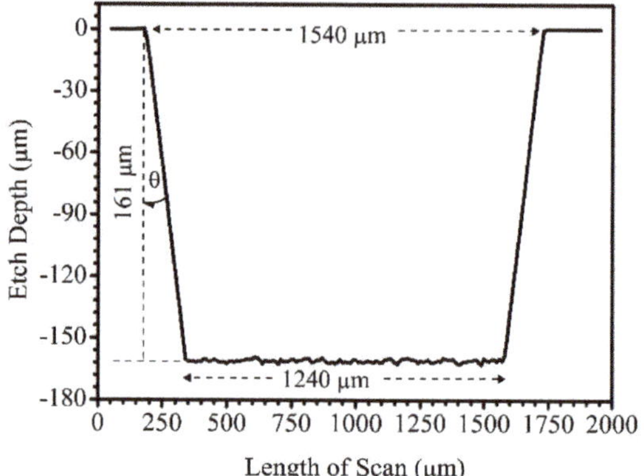

Figure 3. Line profile scan of FLMP etch area (1500 μm × 1500 μm) machined into borosilicate glass substrate at PS_{CNC} of 0.1 mm/s and LF_{av} of 329.06 J/cm^2. The profile was recorded at 5 μm/s, 10 Hz, and 2 mg applied force. The inclined etch surfaces make a contact angle (θ) of ~8° with the vertical plane.

3.2. Etch Depth Dependence on Average Laser Fluence (LF_{av})

The dependence of etch depth on LF_{av} was studied by varying LF_{av} while keeping all other parameters constant. The average laser power was varied from 0.017–1.65 W, which corresponds to LF_{av} values of 14.31–1388.62 J/cm^2, respectively. A linear expression

$$y = 0.1593x + 1.8847 \tag{1}$$

with an excellent R^2 value of 0.991 was obtained as shown in Figure 4. This shows that the etch depth has a strong linear dependence on LF_{av}.

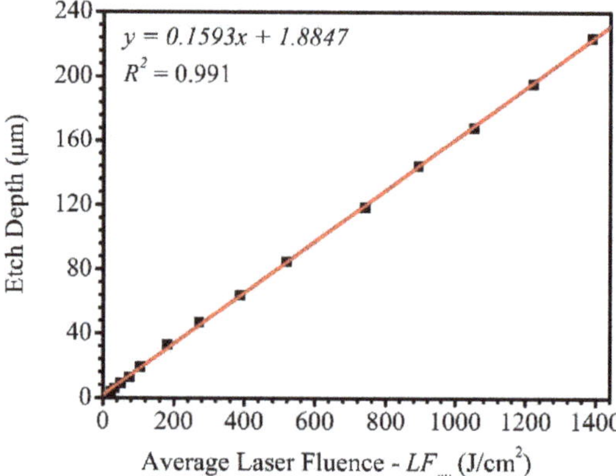

Figure 4. A plot showing the linear dependence of etch depth on average laser fluence (LF_{av}) while keeping all processing parameters constant, such as PS_{CNC} at 0.25 mm/s. Legend—experimental data points: black squares, data fitting: red trace.

The minimum threshold average laser fluence (LF_{av}^{th}) required to etch the borosilicate glass substrate was also investigated. Below 22.72 J/cm^2 (27 μJ), it was found that there was no laser etch on the borosilicate glass substrate at PS_{CNC} of 0.25 mm/s. The PS_{CNC} was further reduced systematically down to 0.025 mm/s, but no laser etch features were observed. Therefore, 22.72 J/cm^2 was determined as the LF_{av}^{th} required for a successful FLMP on borosilicate glass. It must be mentioned that a borosilicate glass with different material composition and specification would have a different LF_{av}^{th}. The corresponding depth at the determined LF_{av}^{th} was 3.9 μm, and this implied that the y-intercept value of 1.8847 μm at 0 J/cm^2 had no physical meaning. This was because the minimum etch depth that could be achieved was 3.87 μm at LF_{av}^{th} = 22.72 J/cm^2, therefore, one could only expect an etch depth of 1.885 μm if the LF_{av}^{th} value was less than 22.72 J/cm^2. Hence, the obtained linear relation was applicable to predict etch depths that were ~≥4 μm deep.

In comparison to the literature, a report by Shin et al. [54] who used FLMP to machine a PDMS substrate obtained a linear relation. The group used 190 fs laser system that produced 343 nm wavelength UV radiation with a maximum average power and pulse energy of 1.8 W and 375 μJ, respectively, at 600 kHz repetition rate. They used a focused laser beam diameter of 5 μm and a relatively fast PS_{CNC} of 500 mm/s, and explored etch depth dependence on increasing LF_{av} in the range of 19.11–382.16 J/cm^2 and obtained a linear relationship. In addition, they studied etch depth dependence on the number of laser beam passes (5–15 multiple passes) on the same surface and observed a similar linear relationship. A similar observation was made by Kam et al. [52] for a silicon wafer substrate. In their study, a 1040 nm wavelength laser with a ~600 fs pulse duration that produced a maximum output power of ~2 W at 200 kHz with a beam spot size of 22 μm diameter was used to machine silicon wafer substrates at 20 μm pitch followed by wet chemical methods. They kept the fluence constant at 3.09 J/cm^2 (9.72 μJ) and increased the number of laser beam passes (multiple pass) as was used by Shin et al. [54]. At constant PS_{CNC} of 30, 120, 480, and 1920 mm/s, a linear relationship for etch depth dependence on multiple number of laser beam passes from ~1–157 was obtained. Though the multiple pass approach is slightly different from increasing the LF_{av} as used in our study, the previous work by Shin et al. [54] has shown that the two methods are comparable as they produce linear relationships for the etch depth dependence.

Also, Crawford et al. investigated etch depth dependence on LF_{av}, and PS_{CNC} by machining linear grooves in a silicon substrate using 800 nm wavelength laser with 150 fs pulse duration [59]. At PS_{CNC} of 0.1–0.5 mm/s, multiple linear relations of etch depths at different LF_{av} regimes were observed. At a lower LF_{av} regime (~< 1.1 J/cm^2), a linear etch depth dependence on fluence was obtained with a slow rise gradient, while at relatively higher LF_{av} regime (~1.1–10 J/cm^2) another linear relationship with a sharp rise gradient was obtained. In comparison to our work on borosilicate glass substrate, the LF_{av}^{th} value of 22.72 J/cm^2 required to observe any etch feature on the substrate was already higher than the highest LF_{av} (10 J/cm^2) investigated by Crawford et al. to etch silicon substrates [59]. However, a similar work by Lee et al. who used 775 nm laser radiation with 150 fs pulse duration to machine silicon wafers over a relatively wide LF_{av} range (< 1000 J/cm^2) also obtained two linear relations for etch depth as a function of LF_{av}. At low (< 10 J/cm^2) and high (10–1000 J/cm^2) LF_{av} regimes, linear logarithmic relationships with slow and fast rise gradients were observed, respectively [48]. These FLMP literature reports, especially the works of Crawford et al. and Lee et al. on silicon wafer substrates strongly support the fact that there are multiple linear etch depth relations, one at low fluence and the other at high fluence. Our study has shown that there is a single etch depth linear dependence on LF_{av} when using FMLP to machine a borosilicate glass substrate. This, to the best of our knowledge, is the first-time experimental determination of such a relation for a borosilicate glass substrate. This is important for future MNT fabrication involving borosilicate glass substrates, such as reservoir micromodels, due to the excellent mechanical strength, exceptional optical transparency, high chemical resistance, and high thermal resistance to the rapid temperature variations of borosilcate glass [60].

3.3. Etch Depth Dependence on CNC Processing Speed (PS_{CNC})

The dependence of etch depth on PS_{CNC} was studied by varying the PS_{CNC} from 0.025–10 mm/s while keeping all other parameters constant, including LF_{av} of 329.06 J/cm^2. All borosilicate glass substrates used in this study were from the same batch unless mentioned otherwise. The data set was fitted to inverse (green trace), logarithm (blue trace), and exponential (red trace) relations as shown in the graph in Figure 5. The fitting results showed that our data agrees more with the exponential plot than the inverse and logarithm relations. This is supported by a better R^2 value of 0.991 for the exponential fitting relative to 0.945 and 0.965 for the inverse and logarithm fitting, respectively. The inverse relation was found to be the worst fitting plot to the experimental data. Here, we obtained a three-phase exponential decay dependence of etch depth on PS_{CNC}. It is observed that the deviation of the inverse fitting curve from the data points increases at $PS_{CNC} > 1$ mm/s.

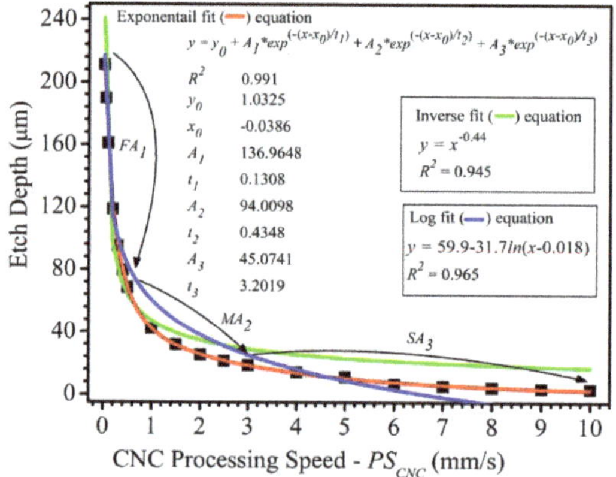

Figure 5. A plot showing an inverse (green trace), logarithmic (blue trace) and a three-phase exponential decay (red trace) dependence of etch depth on CNC processing speed (PS_{CNC}). All processing parameters were kept constant, including average laser fluence (LF_{av}) at 329.06 J/cm^2. Approximate portions of the plot that shows fast, medium, and slow exponential decays are represented by FA_1, MA_2, and SA_3, with pre-exponential decay factors of 136.9648, 94.0098, and 45.0741 µm, respectively. Legend– experimental data points: black squares, inverse data fitting: green trace, logarithmic data fitting: blue trace, exponential data fitting: red trace.

Some literature reports of FLMP on silicon wafer substrates have reported that the etch depth has an inverse dependence on the PS_{CNC}. In the previously discussed work by Crawford et al., an inversely proportional relationship for etch depth (< 25 µm) as a function of PS_{CNC} (0.05–1 mm/s) was obtained when an 800 nm wavelength laser with 150 fs pulse duration was used to machine silicon substrates [59]. A similar observation was made by Kam et al., who explored etch depth (<250 µm) dependence on PS_{CNC} (0.1–1.9 mm/s) by using a 1040 nm wavelength laser with ~600 fs pulse duration to machine silicon wafer substrates [52]. The work of these groups corroborates an earlier work by Ameer-Beg et al., who used a 790 nm wavelength laser with ~170 fs pulse duration to machine fused silica substrate [61]. Ameer-Beg et al. obtained an inversely proportional dependence for etch depth (<40 µm) on PS_{CNC} (1–7 mm/s). It is important to note that most of these literature works found their experimental data obtained for silicon wafer to be in good agreement with an inverse relation, while in our study for borosilicate glass, the inverse relation was the worst to agree with the data.

Also, in the FLMP work of Lee et al. [48], also on silicon wafer substrates that was discussed in Section 3.2, they explored etch depth (<6 µm) dependence on PS_{CNC} (0.5–2.5 mm/s) at a relatively lower LF_{av} range (1.56–6.26 J/cm^2) in comparison to the value of 329.06 J/cm^2 used in this work. They found that etch depth has a one-phase exponential decay dependence on PS_{CNC}. Unlike silicon wafer substrates, there are no such literature studies on borosilicate glass. This is largely due to the widespread use of silicon wafers for MNT fabrication. As previously mentioned in Section 3.2, borosilicate glass is an excellent material for use as reservoir micromodels and microfluidic devices due to its unique material properties such as high optical transparency and high resistance to rapid thermal changes [60]. Hence, it will be important to the MNT community to know borosilicate's fundamental laser-material interaction relationships, such as the etch depth dependence on processing speed. Here, we report the observation of a three-phase exponential decay dependence of etch depth on PS_{CNC} for a borosilicate glass substrate. From regression analysis data fitting, values of 136.965, 94.010, and 45.074 μm were obtained which corresponds to the pre-exponential decay factors for the fast, medium, and slow decay regions, respectively. As shown in Figure 5, there is a good statistical agreement between the experimental data points (black squares) and the three-phase exponential fit (red trace) with an excellent R^2 value of 0.991.

It is worth mentioning that the range of etch depths (3.47–223.8 μm), LF_{av} (14.31–1388.62) and PS_{CNC} (0.025–10 mm/s) investigated in this work is wider than those reported in the literature for commonly used substrates, such as silicon and silica [48,52,59,61]. Pfeiffer et al. have reported on the FLMP of tungsten carbide and steel substrates using 775 nm wavelength radiation with 150 fs pulse duration [53]. Part of their studies explored etch depth dependence on LF_{av} over a total depth range <220 μm, and LF_{av} of 0.2–11 J/cm^2 for the materials. In other studies, polymer substrates such as poly(methyl methacrylate) (PMMA) have been machined with FLMP over etch depth, LF_{av}, and PS_{CNC} of <130 μm, 0.11–1.72 J/cm^2, and 0.5–10 mm/s, respectively [62]. Hence, the wide range of FLMP parameters explored in this study (Table 1) would be applicable and useful as a guide to future FLMP investigations involving many material substrates, including glass, metals, composite materials (e.g., tungsten carbide), and polymers.

3.4. Fabrication of 4 Depth 3D Reservoir Micromodel

A CAD representation of the reservoir micromodel made in Alphacam is shown in Figure 6. The 2D (Figure 6a) and 3D (Figure 6b) top view designs show 3 porous reservoirs (R1, R2, and R3) with inlet channels connected to a common sink. The reservoirs have the same XY dimensions (Figure 6a) but different Z dimensions (depths) as shown in the 3D top view (Figure 6b) and the front view, Figure 6c, that details the various etch depths relative to the substrate surface. The following notations were used; R1 matrix, R1 outer sink, and R1 inner sink that represents the main reservoir matrix, the large outer circular sink, and the small inner circular sink of reservoir 1, etc. Another notation used here is layer 1 and layer 2 which represents the matrices/outer sinks and inner sinks of the reservoirs, respectively. The circular sinks at the bottom of the reservoir matrices have two depths—the large outer circular features have the same depths as their respective R matrices, while the smaller inner circular features have twice as much depth as their respective R matrix depth (Figure 6c). This CAD model shows a total of 4 different depths, i.e., 35 μm (R1), 70 μm (R1 and R2), 140 μm (R2 and R3), and 280 μm (R3) in the same substrate. It is worth mentioning that 6 or more multiple etch depths could have been achieved by using different etch depths for the matrix, outer and inner sinks of each reservoir. However, common etch depths such as 70 μm and 140 μm were used in R1/R2 and R2/R3, respectively, to determine whether etch depths were repeatable among reservoirs which were machined at different times.

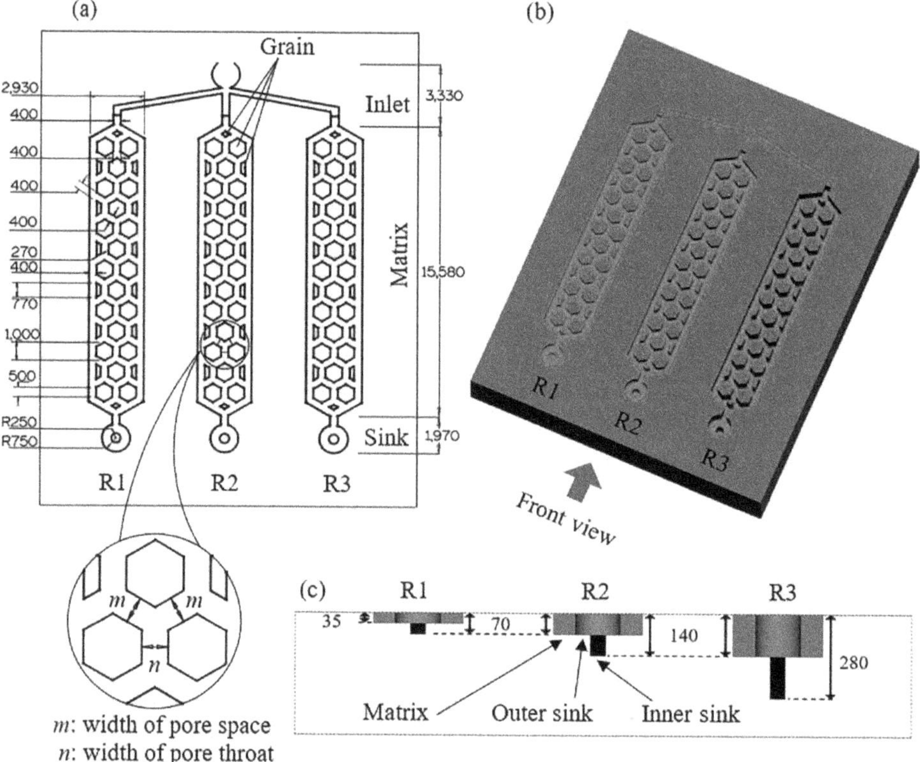

Figure 6. CAD schematic illustration of the 3D multi-depth reservoir (R) micromodel. (**a**) a 2D top view showing a description of all components of the micromodel and their dimensions in μm units, and a zoom-in portion that shows the pore body bounded by 3 solid hexagon grains, and the uniform widths of the pore space (m) and pore throat (n) which gives an aspect ratio ($\frac{m}{n}$) = 1. (**b**) a 3D top view design showing the borosilicate glass substrate (grey) and the etch area (green) with multiple depths. (**c**) a front view of (**b**), grey arrow direction, showing the 4 depths of the reservoir micromodel –35, 70, 140, and 280 μm relative to the surface of the borosilicate substrate. The notations used here are R1 matrix, R1 outer sink, and R1 inner sink that represents the main reservoir matrix, the large outer circular sink, and the small inner circular sink of reservoir 1, etc. Emphasis was placed on the reservoir matrices and sinks, therefore portions (grey area) of the inlet channel was not etched as shown in (**b**).

Pore dimensions, such as size and shape, are known to influence fluid flow in porous media [9,10,63–65]. Here, the pore body is bounded by 3 solid hexagon grains as shown in the zoom-in inset of Figure 6a. The pore space, m, longest distance between two solid grains, and pore throat, n, shortest distance between two solid grains [9], have uniform width of 400 μm, producing an aspect ratio $\frac{m}{n} = 1$. The solid grains in the reservoir matrix are mainly composed of large hexagons and small trapezoid geometries. The dimensions of the hexagons were 1000 μm (length) and 800 μm (breath) which gives an aspect ratio of 1.25. Similarly, the length and breadth of the trapezoid grains were 650 and 200 μm, respectively, producing an aspect ratio of 3.25. Each reservoir had a total surface area and etch surface area of 6.43×10^7 and 4.99×10^7 μm^2, respectively, which results in a surface porosity of 77.6%. The different depths of R1, R2 and R3 produced total etch volumes of 2.25×10^9, 4.82×10^9, and 9.00×10^9 μm^3, respectively. Prior to machining the inner sinks of R1, R2, and R3, the laser beam was refocused at the newly etched surface of layer 1 by moving down the vertical Z axis by 35, 70, and 140 μm, respectively.

3.5. Calibration Curves as Models to Predict FLMP Parameters

Here, the calibration curves produced in Sections 3.2 and 3.3 were used as models to predict the processing parameters required to fabricate the 3D multi-depth reservoir micromodel. This afforded us the ability to test the accuracy of our model and the FLMP method. From the CAD in Figure 6, the reservoirs—R1, R2, and R3, have the same XY dimensions but different depths of 35, 70, and 140 μm for the reservoir matrices/outer sinks, and a total depth of 70, 140, and 280 μm for the inner circular sinks, respectively. The etch depth dependence on LF_{av} calibration curve requires that all FLMP parameters be kept constant while varying LF_{av} to achieve the required etch depth. Alternatively, the etch depth dependence on PS_{CNC} calibration curve was used due to ease of control of PS_{CNC} in comparison to LF_{av} in our experimental setup. The predicted PS_{CNC} necessary to achieve the desired etch depths across the reservoir micromodel are given in Table 2 and will be discussed later.

Table 2. Surface profilometer depth characterization of the 3D multi-depth reservoir micromodel machined into borosilicate glass substrate using FLMP. The experimental machined etch depths are compared to model predictions. The depths of layer 2 are relative to the etched surface of layer 1. NB: (-) % error indicates that the machined etch depth value was < values predicted by the calibration model.

Reservoirs		CAD Etch Depth (μm)	Etch Depth vs. PS_{CNC} Model Prediction (mm/s)	Experimental Etch Depth (μm)			Average ± σ (μm)	%Error
Layer 1	R1 Matrix	35.0	1.301	38.4	34.8	35.7	36.3 ± 1.9	3.7
	R2 Matrix	70.0	0.485	71.6	69.3	69.2	70.0 ± 1.4	0.0
	R3 Matrix	140.0	0.142	141.2	137.6	141.0	140.0 ± 2.0	0.0
	R1 Outer Sink	35.0	1.301	33.0	33.2	-	33.1 ± 0.1	−5.4
	R2 Outer Sink	70.0	0.485	72.5	72.1	-	72.3 ± 0.3	3.3
	R3 Outer Sink	140.0	0.142	152.3	152.1	-	152.2 ± 0.1	8.7
Layer 2	R1 Inner Sink	35.0	1.301	36.1	-	-	36.1	3.1
	R2 Inner Sink	70.0	0.485	65.2	-	-	65.2	−6.9
	R3 Inner Sink	140.0	0.142	128.9	-	-	128.9	−7.9

3.6. Characterization of 3D Multi-Depth Reservoir Micromodel in Borosilicate Glass

Images of the 3D multi-depth reservoir micromodel machined in borosilicate glass using FLMP are shown in Figure 7. The overview of the micromodel (Figure 7a) covers approximately 20 mm × 15 mm surface. It took ~10 h to machine all the various components and depths of the micromodel. Figure 7b shows a portion of the R2 matrix that highlights the solid grains, i.e., hexagon and trapezoid geometries, that are separated from each other by a homogenous micro channel network. The red line indicates the surface profilometer path used to scan the etch depths for R1, R2, and R2 matrices. A continuous line profile scan across each reservoir, as shown by the red line, produces three depth measurements for each R that was expected to be equal. A similar approach was used to measure the etch depths across the circular sinks. Figure 7c–e show the respective portions of R1, R2, and R3 inlet channels. Here, it is shown that FLMP can make both sharp and curved etch features unlike wet photolithography methods that produce curved/rounded features as reported by others [13,66].

Through visual inspection, portions of the R1 inlet channel showed high surface roughness. The average surface roughness, *Ra*, was measured at multiple locations across the etched surface of the reservoirs and resulted in *Ra* values of 525, 320, and 800 nm for R1, R2, and R3, respectively. The *Ra* for the unmachined glass substrate was ~0.5 nm. Each value was obtained by averaging 5 experimental measurements. This *Ra* data set is not enough to predict a meaningful relationship, such as the dependence of *Ra* on etch depth and/or *Ra* on PS_{CNC}. Details of this comprehensive investigation will be presented in another manuscript as previously mentioned. However, it is worth mentioning that the highest *Ra* value of 800 nm obtained for R3, which has the deepest etch depth relative to R1 and R2, agrees with reports by others [48,52]. At constant LF_{av}, deeper channels (e.g., R3 with 140 μm etch depth) are obtained at slower PS_{CNC} (0.142 mm/s) in comparison to etch depths of R2 (70 µm) or R1 (35 µm) that were machined at higher speeds of 0.485 and 1.301 mm/s, respectively. The slower PS_{CNC} increases thermal effects due to proximity and overlap of laser pulses that is accompanied by debris build up. The debris occupies the channels, blocking the laser beam to the desired target surfaces which increases the surface roughness. The roughness of the unetched glass surfaces right next to the etched structures was also determined by collecting surface profilometer scans at multiple locations, including the inlet, matrix, and sinks of all 3 reservoirs. Each data point was recorded near the etched structures by a 100 µm long scan. More than 20 data points were averaged to produce an *Ra* of 3.7 nm with a wide deviation of ±5.1 nm. This range of roughness (3.7 ± 5.1 nm) around the etched features compares reasonably well to the furthest (> 2000 µm) unetched area roughness of ~0.5 nm. This should not pose challenges for applications requiring bonding of a lid to the top of the borosilicate glass substrate to form a sealed channel or chamber.

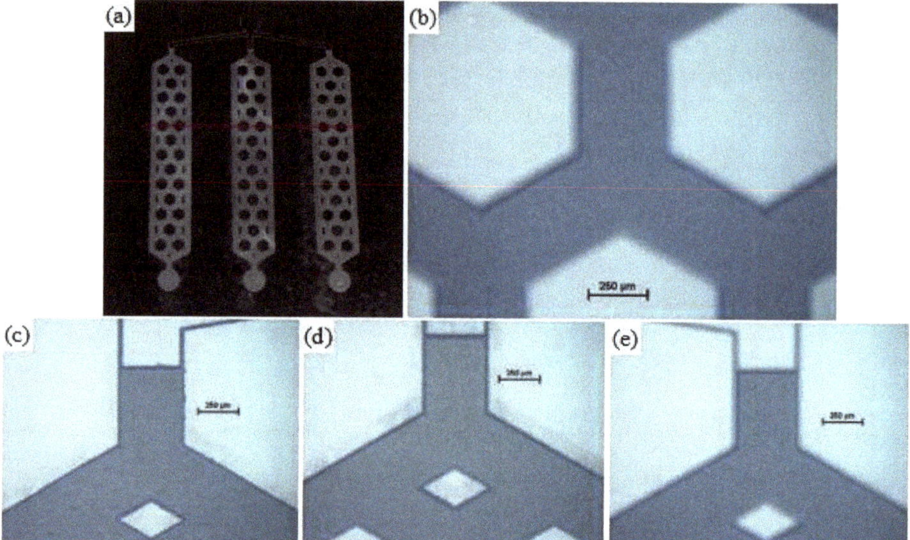

Figure 7. Images of several sections of the FLMP fabricated 3D multi-depth reservoir micromodel: (**a**) the red line illustrates the line profile scan path which goes through three pore spaces and two hexagonal pore bodies for each reservoir (R), (**b**) a zoom-in section of reservoir 3 where the image was focused at the etched surface, and (**c**–**e**) shows inlet portions of reservoirs 1, 2, and 3, respectively. Images (**b**,**e**) looks blurrier than (**c**,**d**) due to deeper depth as the microscope was focused on the bases of the channels. The scale bars are 250 µm.

Figure 8 shows a line profile scan across the reservoir matrices. The black, red, and blue traces represent the line profile scans across the matrices of R1, R2, and R3, respectively. The line profile scans are vertically stacked up in the graph, and this illustrates how neatly all the etch profiles overlap across the matrices of all three reservoirs. This agrees with the etch profile shown in Figure 3. Also, it shows the robustness of the FLMP technique which makes it possible to use calibration curves as models to predict experimental parameters required for future experiments. The measured etch depths, including standard deviations, and percentage errors are given in Table 2 above. The average etch depths for the matrices of R1, R2, and R3 were 36.3, 70.0, and 140.0 µm in comparison to the model prediction values of 35, 70.0, and 140.0 µm, which produces percentage errors of 3.7, 0.0, and 0.0%, respectively. This shows that our approach of using FLMP technique to machine 3D multi-depth features has good accuracy in producing the required results predicted by the calibration model.

Also, the results in Table 2 shows standard deviations (σ) of ≤ 2 μm, indicating a good repeatability of etch depths across the large etch surface of all three reservoirs. Optical microscope and surface profilometer images of the sinks of R3 and R2 are shown in Figure 9. The images of Figures 9a and 9b, were recorded consecutively by focusing the microscope at the etch surfaces of the outer and inner sinks of R3, respectively. Some remaining debris can be seen in both images at the lower portion of the inner sink; a common occurrence observed by other investigators in deep channels and pockets machined by FLMP [48,52,56,59]. This indicates that >30 min of sonication in isopropyl alcohol bath was probably required to remove all remaining debris. The 3D image of reservoir R2 sink in Figure 9c provides a complementary visual observation to the microscope images. It emphasizes the vertical depth information where the 2D microscope is lacking. The recorded total etch depth of 144 μm in Figure 9c differs marginally by ~6 μm in comparison to the combined 2D line profile etch depths of the outer and inner sinks of R2 (138 μm).

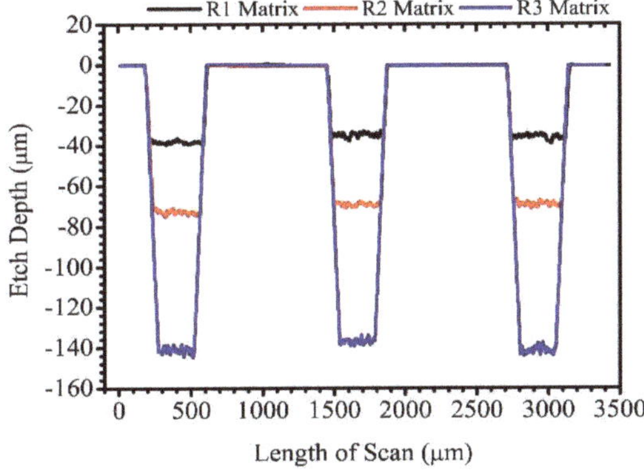

Figure 8. A plot showing the surface profilometer line scans across the matrices of the 3D multi-depth reservoir micromodel shown in Figure 7a. The black, red, and blue traces correspond to the profiles of reservoirs R1, R2, and R3, respectively.

A graph showing 2D line profile scans across the sinks of R1, R2, and R3 is shown in Figure 10 and their measured etch depths are given in Table 2, showing the different etch depths. The instrument limit of the surface profilometer was ~270 μm which was observed by a linear and smooth horizontal etch surface (blue trace arrow) in R3. Therefore, another profile scan was done by starting from the etched layer 1 surface of R3 (green trace). Here also, a good overlap was observed between the two profiles (blue and green traces).

The average experimental etch depth values obtained for the outer sinks of R1, R2, and R3 were 33.1 ± 0.1, 72.3 ± 0.3, and 152.2 ± 0.1 μm with percentage errors of −5.4, 3.3, and 8.7% in comparison to the calibration model prediction values of 35.0, 70.0, and 140.0 μm, respectively. The smaller deviations (σ ≤ 0.3 μm) observed for the outer sinks in comparison to the matrices (σ ≤ 2.0 μm) of the reservoirs could be due to localized etching in the former than the latter. For example, the experimental etch depth values of 33.0 and 33.2 μm for R1 outer sink were machined in a relatively small surface area while that of the matrix (38.4, 34.8, and 35.7 μm) was spread over a large etch area as shown in Figures 7 and 9. This would allow small variations in the substrate, such as material density and surface height fluctuations, to slightly impact the resultant etch depths.

A similar range of percentage errors was obtained for the inner sinks (3.1–7.9%) relative to the outer sinks (3.3–8.7%) as shown in Table 2. Here, experimental etch depth values of 36.1, 65.2, and 128.9 μm were obtained for the inner sinks of R1, R2, and R3 with respective percentage errors of 3.1, −6.9, and −7.9%. Generally, it was observed that etch depths in layer 2 (i.e., inner sinks) were shallower than the values predicted by the model while those in layer 1 (matrices and outer sinks) had deeper depths.

The etch depths of R2 (65.2 μm) and R3 (128.9 μm) inner sinks were shallower than their respective model predicted values of 70.0 and 140.0 μm, except R1 inner sink (36.1 μm) which was deeper than the predicted value of 35.0 μm. On the contrary, most of the etch depths in layer 1—i.e., R1, R2 and R3 matrices, and R2 and R3 outer sinks, except R1 outer sink, were deeper than their respective values predicted by the calibration model as shown in Table 2. This was largely attributed to remaining debris on the surface of layer 1 that partially impedes the laser beam from direct interaction with the etched surface during the machining of layer 2 features as previously discussed above and reported in the literature by others [48,52,56,59]. The remaining debris on the layer 1 etched surface competes with material removal in layer 2, slightly impacting the efficiency of the etching process which results in reduction in predicted etch depths as observed here. Also, another reason for observing shallower etch depths in deeper channels is the limitation on etch volume due to the beams focal volume at constant Z position (focal distance)—i.e., the volume of material removed decreases significantly beyond the region where the fluence is tightly focused [39,44]. However, in Section 3.2, etch depths of 251.1 and 270.7 μm, which are deeper than the predicted values of R2 (70.0 μm) and R3 (140.0 μm) inner sinks, were successfully obtained due to their low aspect ratios (<0.2) relative to that of the inner sinks (<0.4). Therefore, the major factor responsible for the shallower than predicted etch depths for most of the layer 2 features is largely attributed to the impedance by the remaining debris on layer 1 to the laser beam.

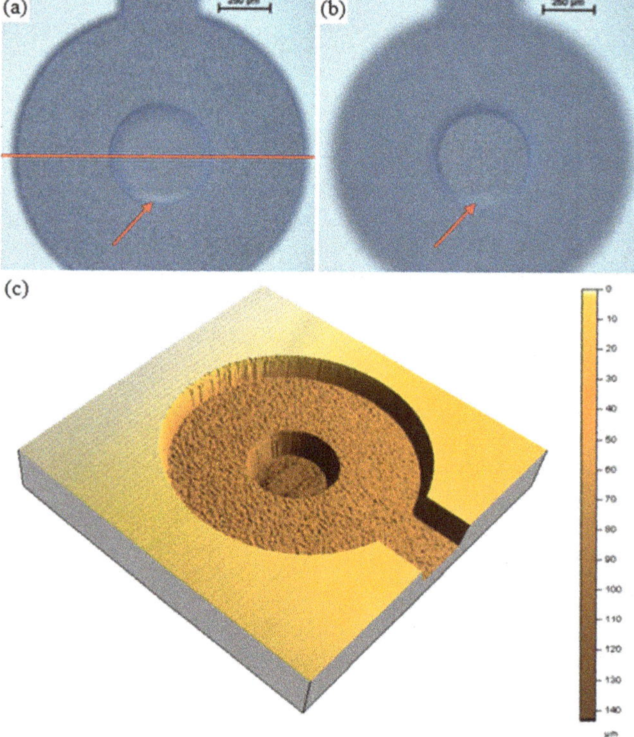

Figure 9. Images of the 3D multi-depth reservoir micromodel (**a**,**b**) taken with an optical microscope and showing the (**a**) outer and (**b**) inner circular sinks of R3 in focus. The red line across (**a**) indicates a 2D line profile scan path while (**c**) is a 3D image of reservoir R2 sink recorded with the surface profilometer at 2 mg force, 10 μm/s speed, 10 Hz, and 3 μm scan interval. The arrows in (**a**,**b**) point to the location of machined debris remaining after sonication.

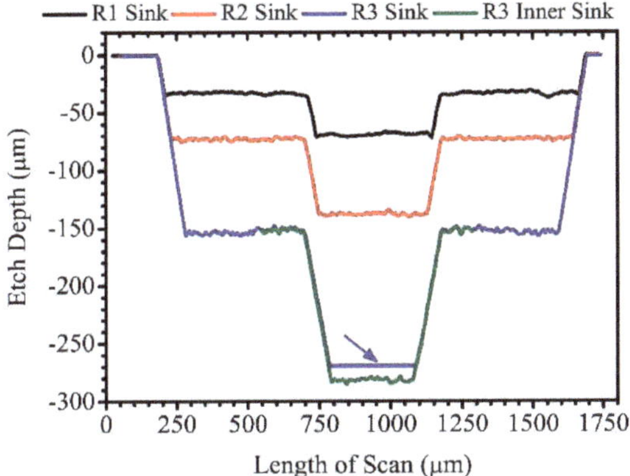

Figure 10. A plot of line profile scans across the circular sinks of the 3D multi-depth reservoir micromodel. The black, red, and blue traces correspond to the profiles of reservoirs R1, R2, and R3, respectively. The green trace was a repeated scan for reservoir R3 from the newly etched outer sink surface (layer 1) due to instrument limit which is observed as a smooth horizontal etch surface (indicated by blue arrow).

4. Conclusions

The fabrication of a 4-depth 3D reservoir micromodel over a large surface area (20 mm × 15 mm) in a borosilicate glass substrate has been reported, for the first time, using femtosecond laser material processing (FLMP). The etch profile of the focused laser beam showed a Gaussian-like profile that makes ~8° contact angle with the vertical plane. The dependence of etch depth on two major FLMP parameters – average laser fluence (LF_{av}), and CNC processing speed (PS_{CNC}), was studied. It was found that etch depth has a strong linear dependence on LF_{av} with an excellent R^2 value of 0.991. Also, a threshold average laser fluence (LF_{av}^{th}) value of 22.72 J/cm^2 was determined as the minimum energy density required to etch borosilicate glass. The experimental data points for etch depth dependence on PS_{CNC} was fitted to an inverse, logarithm, and exponential relations. Our data for borosilicate glass was in better agreement with the exponential relation, while other substrates such as silicon, have been shown in the literature to agree more with an inverse relation. It was shown that etch depth has a three-phase exponential decay dependence on PS_{CNC}, with another excellent R^2 value of 0.991. The linear and three-phase exponential decay relationships were successfully used as models to predict processing parameters required to machine the 3D multi-depth reservoir micromodel.

The etch depth dependence on PS_{CNC} model was used to machine the 3D reservoir micromodel composed of 4 etch depths, i.e., 35, 70, 140, and 280 μm, in the same borosilicate glass substrate. The experimental etch depths showed good results accuracy with percentage errors ≤ 8.7% in comparison to the model prediction values. Deviations of ≤ 2.0 μm in depth were achieved which showed that the etch depths were repeatable across the large etched surface of the 3-reservoir micromodel consisting of 4 multiple depths. Thus, this study has shown the robustness of FLMP as a fabrication technique to produce reliable etch depth results across a large surface area in a borosilicate glass substrate. In addition, it was shown that the etch depth dependency models produced in this study will be useful to guide the work of future researchers. This study will help the development and fabrication of micro/nanotechnology (MNT) systems, including microfluidic devices that are used for transport investigations in porous media.

Author Contributions: Conceptualization, E.O.-A. and C.D.; Methodology, E.O.-A. and C.D.; Resources, C.D.; Software, E.O.-A.; Validation, E.O.-A.; Formal analysis, E.O.-A.; Investigation, E.O.-A.; Data Curation, E.O.-A.; Writing—Original draft preparation, E.O.-A.; Writing—Review and editing, C.D.; Project administration, C.D.;

Visualization, E.O.-A.; Supervision, C.D.; and Funding acquisition, E.O.-A. and C.D. All authors have read and agreed to the published version of the manuscript.

Funding: This research was funded by CMC Microsystems (PDES-RD156), and Natural Sciences and the Engineering Research Council of Canada (NSERC).

Acknowledgments: The authors would like to acknowledge CMC Microsystems for the provision of products and services that facilitated this research, including CAD tools such as Alphacam. We thank Dylan Finch and Malcom Haynes for their help with calibration and maintenance of measurement equipment at the Microsystems Hub, University of Calgary.

Conflicts of Interest: The authors declare no conflict of interest.

References

1. McKellar, M.; Wardlaw, N.C. A Method of Making Two-Dimensional Glass Micromodels of Pore Systems. *J. Can. Pet. Technol.* **1982**, *21*, 39–41. [CrossRef]
2. Lago, M.; Huerta, M.; Gomes, R. Visualization Study during Depletion Experiments of Venezuelan Heavy Oils Using Glass Micromodels. *J. Can. Pet. Technol.* **2002**, *41*, 41–47. [CrossRef]
3. Mejia, L.; Tagavifar, M.; Xu, K.; Mejia, M.; Du, Y.; Balhoff, M. Surfactant Flooding in Oil-Wet Micromodels with High Permeability Fractures. *Fuel* **2019**, *241*, 1117–1128. [CrossRef]
4. Lyu, X.; Liu, H.; Pang, Z.; Sun, Z. Visualized Study of Thermochemistry Assisted Steam Flooding to Improve Oil Recovery in Heavy Oil Reservoir with Glass Micromodels. *Fuel* **2018**, *218*, 118–126. [CrossRef]
5. Yun, W.; Chang, S.; Cogswell, D.A.; Eichmann, S.L.; Gizzatov, A.; Thomas, G.; Al-Hazza, N.; Abdel-Fattah, A.; Wang, W. Toward Reservoir-on-a-Chip: Rapid Performance Evaluation of Enhanced Oil Recovery Surfactants for Carbonate Reservoirs Using A Calcite-Coated Micromodel. *Sci. Rep.* **2020**, *10*, 782. [CrossRef] [PubMed]
6. Chalbaud, C.; Robin, M.; Lombard, J.-M.; Bertin, H.; Egermann, P. Brine/CO2 Interfacial Properties and Effects on CO2 Storage in Deep Saline Aquifers. *Oil Gas Sci. Technol. Rev. d'IFP Energies Nouv.* **2010**, *65*, 541–555. [CrossRef]
7. Morais, S.; Liu, N.; Diouf, A.; Bernard, D.; Lecoutre, C.; Garrabos, Y.; Marre, S. Monitoring CO2 Invasion Processes at the Pore Scale Using Geological Labs on Chip. *Lab Chip* **2016**, *16*, 3493–3502. [CrossRef]
8. Song, W.; de Haas, T.W.; Fadaei, H.; Sinton, D. Chip-off-the-Old-Rock: The Study of Reservoir-Relevant Geological Processes with Real-Rock Micromodels. *Lab Chip* **2014**, *14*, 4382–4390. [CrossRef]
9. Zhang, J.; Zhang, H.; Lee, D.; Ryu, S.; Kim, S. Microfluidic Study on the Two-Phase Fluid Flow in Porous Media during Repetitive Drainage-Imbibition Cycles and Implications to the CAES Operation. *Transp. Porous Media* **2020**, *131*, 449–472. [CrossRef]
10. Seyyedi, M.; Giwelli, A.; White, C.; Esteban, L.; Verrall, M.; Clennell, B. Effects of Geochemical Reactions on Multi-Phase Flow in Porous Media during CO2 Injection. *Fuel* **2020**, *269*, 117421. [CrossRef]
11. Sinton, D. Energy: The Microfluidic Frontier. *Lab Chip* **2014**, *14*, 3127–3134. [CrossRef] [PubMed]
12. Tagavifar, M.; Xu, K.; Jang, S.H.; Balhoff, M.T.; Pope, G.A. Spontaneous and Flow-Driven Interfacial Phase Change: Dynamics of Microemulsion Formation at the Pore Scale. *Langmuir* **2017**, *33*, 13077–13086. [CrossRef] [PubMed]
13. Conn, C.A.; Ma, K.; Hirasaki, G.J.; Biswal, S.L. Visualizing Oil Displacement with Foam in A Microfluidic Device with Permeability Contrast. *Lab Chip* **2014**, *14*, 3968–3977. [CrossRef] [PubMed]
14. Rossen, W.R. Comment on "Verification of Roof Snap-off as a Foam-Generation Mechanism in Porous Media at Steady State". *Colloids Surf. A Physicochem. Eng. Asp.* **2008**, *322*, 261–269. [CrossRef]
15. Krummel, A.T.; Datta, S.S.; Münster, S.; Weitz, D.A. Visualizing Multiphase Flow and Trapped Fluid Configurations in A Model Three-Dimensional Porous Medium. *AIChE J.* **2013**, *59*, 1022–1029. [CrossRef]
16. Xu, K.; Liang, T.; Zhu, P.; Qi, P.; Lu, J.; Huh, C.; Balhoff, M. A 2.5-D Glass Micromodel for Investigation of Multi-Phase Flow in Porous Media. *Lab Chip* **2017**, *17*, 640–646. [CrossRef] [PubMed]
17. Anbari, A.; Chien, H.-T.; Datta, S.S.; Deng, W.; Weitz, D.A.; Fan, J. Microfluidic Model Porous Media: Fabrication and Applications. *Small Nano Micro.* **2018**, *14*, 1703575. [CrossRef] [PubMed]
18. Chrysikopoulos, C.V.; Plega, C.C.; Katzourakis, V.E. Non-Invasive in Situ Concentration Determination of Fluorescent or Color Tracers and Pollutants in A Glass Pore Network Model. *J. Hazard. Mater.* **2011**, *198*, 299–306. [CrossRef]

19. Yun, W.; Ross, C.M.; Roman, S.; Kovscek, A.R. Creation of A Dual-Porosity and Dual-Depth Micromodel for the Study of Multiphase Flow in Complex Porous Media. *Lab Chip* **2017**, *17*, 1462–1474. [CrossRef]
20. Karadimitriou, N.K.; Joekar-Niasar, V.; Hassanizadeh, S.M.; Kleingeld, P.J.; Pyrak-Nolte, L.J. A Novel Deep Reactive Ion Etched (DRIE) Glass Micro-Model for Two-Phase Flow Experiments. *Lab Chip* **2012**, *12*, 3413–3418. [CrossRef]
21. Thienot, E.; Domingo, F.; Cambril, E.; Gosse, C. Reactive Ion Etching of Glass for Biochip Applications: Composition Effects and Surface Damages. *Microelectron. Eng.* **2006**, *83*, 1155–1158. [CrossRef]
22. Wu, B.; Kumar, A.; Pamarthy, S. High Aspect Ratio Silicon Etch: A Review. *J. Appl. Phys.* **2010**, *108*, 51101. [CrossRef]
23. Karadimitriou, N.K.; Hassanizadeh, S.M. A Review of Micromodels and Their Use in Two-Phase Flow Studies. *Vadose Zo. J.* **2012**, *11*, vzj2011.0072. [CrossRef]
24. Au, A.K.; Huynh, W.; Horowitz, L.F.; Folch, A. 3D-Printed Microfluidics. *Angew. Chem. Int. Ed.* **2016**, *55*, 3862–3881. [CrossRef] [PubMed]
25. Seers, T.D.; Alyafei, N. Open Source Toolkit for Micro-Model Generation Using 3D Printing. In *SPE Europec Featured at 80th EAGE Conference and Exhibition, Copenhagen, Denmark*; Society of Petroleum Engineers: Copenhagen, Denmark, 2018; p. SPE-190852-MS.
26. Giridhar, M.S.; Seong, K.; Schülzgen, A.; Khulbe, P.; Peyghambarian, N.; Mansuripur, M. Femtosecond Pulsed Laser Micromachining of Glass Substrates with Application to Microfluidic Devices. *Appl. Opt.* **2004**, *43*, 4584–4589. [CrossRef] [PubMed]
27. Sugioka, K.; Cheng, Y.; Midorikawa, K. Three-Dimensional Micromachining of Glass Using Femtosecond Laser for Lab-on-a-Chip Device Manufacture. *Appl. Phys. A* **2005**, *81*, 1–10. [CrossRef]
28. Hayden, C.J.; Dalton, C. Direct Patterning of Microelectrode Arrays Using Femtosecond Laser Micromachining. *Appl. Surf. Sci.* **2010**, *256*, 3761–3766. [CrossRef]
29. Italia, V.; Giakoumaki, A.N.; Bonfadini, S.; Bharadwaj, V.; Le Phu, T.; Eaton, S.M.; Ramponi, R.; Bergamini, G.; Lanzani, G.; Criante, L. Laser-Inscribed Glass Microfluidic Device for Non-Mixing Flow of Miscible Solvents. *Micromachines* **2018**, *10*, 23. [CrossRef]
30. Fulton, A.L.; Beebe, D.J.; Sackmann, E.K.; Fulton, A.L.; Beebe, D.J. The Present and Future Role of Microfluidics in Biomedical Research. *Nature* **2014**, *507*, 181–189.
31. Nguyen, N.-T.; Wu, Z. Micromixers—A Review. *J. Micromech. Microeng.* **2005**, *15*, R1–R16. [CrossRef]
32. Chang, T.-C.; Wang, S.-C.; Chien, C.-W.; Cheng, C.-W.; Lee, C.-Y. Using Femtosecond Laser to Fabricate the Interior 3D Structures of Polymeric Microfluidic Biochips. *J. Laser Micro Nanoeng. JLMN* **2011**, *6*, 87–90.
33. Dalton, C.; Hayden, C.J.; Burt, J.P.; Manz, A.; Eijkel, J.C.T.; Burt, J.P.H. A Circular Ac Magnetohydrodynamic Micropump for Chromatographic Applications. *Sens. Actuators B Chem.* **2003**, *92*, 215–221.
34. Voldman, J.; Jaffe, A. Multi-Frequency Dielectrophoretic Characterization of Single Cells. *Microsyst. Nanoeng.* **2018**, *4*, 23.
35. Khalil, A.A.; Lalanne, P.; Bérubé, J.-P.; Petit, Y.; Vallée, R.; Canioni, L. Femtosecond Laser Writing of Near-Surface Waveguides for Refractive-Index Sensing. *Opt. Express* **2019**, *27*, 31130–31143. [CrossRef] [PubMed]
36. Balaji, V.; Castro, K.; Folch, A.; Balaji, V.; Castro, K.; Folch, A. A Laser-Engraving Technique for Portable Micropneumatic Oscillators. *Micromachines* **2018**, *9*, 426. [CrossRef] [PubMed]
37. Chen, G.Y.; Piantedosi, F.; Otten, D.; Kang, Y.Q.; Zhang, W.Q.; Zhou, X.; Monro, T.M.; Lancaster, D.G. Femtosecond-Laser-Written Microstructured Waveguides in BK7 Glass. *Sci. Rep.* **2018**, *8*, 10377. [CrossRef] [PubMed]
38. Owusu-Ansah, E.; Dalton, C.; Cully, C. Femtosecond Laser Machining of Complex X-Ray Masks in Tungsten Sheets. Unpublished work.
39. Phillips, K.C.; Gandhi, H.H.; Mazur, E.; Sundaram, S.K. Ultrafast Laser Processing of Materials: A Review. *Adv. Opt. Photonics* **2015**, *7*, 684–712. [CrossRef]
40. De Angelis, R.; Duvillaret, L.; Andreoli, P.L.; Cipriani, M.; Consoli, F.; De Angelis, R.; Duvillaret, L.; Andreoli, P.L.; Cipriani, M.; Cristofari, G.; et al. Time-Resolved Absolute Measurements by Electro-Optic Effect of Giant Electromagnetic Pulses Due to Laser-Plasma Interaction in Nanosecond Regime. *Sci. Rep.* **2016**, *6*, 27889.

41. Owusu-Ansah, E.; Horwood, C.A.; El-Sayed, H.A.; Birss, V.I.; Shi, Y.J. A Method for the Formation of Pt Metal Nanoparticle Arrays Using Nanosecond Pulsed Laser Dewetting. *Appl. Phys. Lett.* **2015**, *106*, 203103. [CrossRef]
42. Owusu-Ansah, E.; Birss, V.I.; Shi, Y. Mechanisms of Pulsed Laser-Induced Dewetting of Thin Platinum Films on Tantalum Substrates—A Quantitative Study. *J. Phys. Chem. C* **2020**, *124*, 23387–23393. [CrossRef]
43. Winkler, M.T. Non-Equilbrium Chalcogen Concentrations in Silicon: Physical Structure, Electronic Transport, and Photovoltaic Potential. PhD Thesis, Harvard University, Cambridge, MA, USA, 2009.
44. Sugioka, K.; Cheng, Y. Femtosecond Laser Three-Dimensional Micro- and Nanofabrication. *Appl. Phys. Rev.* **2014**, *1*, 41303. [CrossRef]
45. Garrido-Diez, D.; Baraia, I. Review of Wide Bandgap Materials and Their Impact in New Power Devices. In *2017 IEEE International Workshop of Electronics, Control, Measurement, Signals and their Application to Mechatronics (ECMSM)*; IEEE: Piscataway, NJ, USA, 2017; pp. 1–6.
46. Ben-Yakar, A.; Byer, R.L. Femtosecond Laser Ablation Properties of Borosilicate Glass. *J. Appl. Phys.* **2004**, *96*, 5316–5323. [CrossRef]
47. Hayden, C.J. A Simple Three-Dimensional Computer Simulation Tool for Predicting Femtosecond Laser Micromachined Structures. *J. Micromech. Microeng.* **2010**, *20*, 25010. [CrossRef]
48. Lee, S.; Yang, D.; Nikumb, S. Femtosecond Laser Micromilling of Si Wafers. *Appl. Surf. Sci.* **2008**, *254*, 2996–3005. [CrossRef]
49. Li, C.; Nikumb, S.; Wong, F. An Optimal Process of Femtosecond Laser Cutting of NiTi Shape Memory Alloy for Fabrication of Miniature Devices. *Opt. Lasers Eng.* **2006**, *44*, 1078–1087. [CrossRef]
50. Grojo, D. Internal Structuring of Silicon by Ultrafast Laser Irradiation. In *2019 Conference on Lasers and Electro-Optics Europe and European Quantum Electronics Conference*; Optical Society of America: Washington, DC, USA, 2019; p. cm_2_1.
51. Stuart, B.C.; Feit, M.D.; Herman, S.; Rubenchik, A.M.; Shore, B.W.; Perry, M.D. Optical Ablation by High-Power Short-Pulse Lasers. *J. Opt. Soc. Am. B* **1996**, *13*, 459–468. [CrossRef]
52. Shah, L.; Mazumder, J.; Kam, D.H.; Shah, L.; Mazumder, J. Femtosecond Laser Machining of Multi-Depth Microchannel Networks onto Silicon. *J. Micromech. Microeng.* **2011**, *21*, 45027.
53. Pfeiffer, M.; Engel, A.; Weissmantel, S.; Scholze, S.; Reisse, G. Microstructuring of Steel and Hard Metal Using Femtosecond Laser Pulses. In *Lasers in Manufacturing 2011: Proceedings of the Sixth International WLT Conference on Lasers in Manufacturing*; Schmidt, M., Zaeh, M.F., Graf, T., Ostendorf, A., Eds.; Physics Procedia; Elsevier Science BV: Amsterdam, The Netherlands, 2011; Volume 12 Pt B, pp. 60–66.
54. Shin, H.; Kim, H.; Jang, Y.; Jung, J.; Oh, J. Femtosecond Laser-Inscripted Direct Ultrafast Fabrication of A DNA Distributor Using Microfluidics. *Appl. Sci.* **2017**, *7*, 1083. [CrossRef]
55. Lei, C.; Pan, Z.; Jianxiong, C.; Tu, P. Influence of Processing Parameters on the Structure Size of Microchannel Processed by Femtosecond Laser. *Opt. Laser Technol.* **2018**, *106*, 47–51. [CrossRef]
56. Huang, Y.; Wu, X.; Liu, H.; Jiang, H. Fabrication of Through-Wafer 3D Microfluidics in Silicon Carbide Using Femtosecond Laser. *J. Micromech. Microeng.* **2017**, *27*, 65005. [CrossRef]
57. Ito, Y.; Yoshizaki, R.; Miyamoto, N.; Sugita, N. Ultrafast and Precision Drilling of Glass by Selective Absorption of Fiber-Laser Pulse into Femtosecond-Laser-Induced Filament. *Appl. Phys. Lett.* **2018**, *113*, 61101. [CrossRef]
58. Chalupský, J.; Krzywinski, J.; Juha, L.; Hájková, V.; Cihelka, J.; Burian, T.; Vyšín, L.; Gaudin, J.; Gleeson, A.; Jurek, M.; et al. Spot Size Characterization of Focused Non-Gaussian X-Ray Laser Beams. *Opt. Express* **2010**, *18*, 27836–27845. [CrossRef] [PubMed]
59. Crawford, T.H.R.; Borowiec, A.; Haugen, H.K. Femtosecond Laser Micromachining of Grooves in Silicon with 800 Nm Pulses. *Appl. Phys. A* **2005**, *80*, 1717–1724. [CrossRef]
60. Zamuruyev, K.O.; Zrodnikov, Y.; Davis, C.E. Photolithography-Free Laser-Patterned {HF} Acid-Resistant Chromium-Polyimide Mask for Rapid Fabrication of Microfluidic Systems in Glass. *J. Micromech. Microeng.* **2016**, *27*, 15010. [CrossRef] [PubMed]
61. Ameer-Beg, S.; Perrie, W.; Rathbone, S.; Wright, J.; Weaver, W.; Champoux, H. Femtosecond Laser Microstructuring of Materials. *Appl. Surf. Sci.* **1998**, *127–129*, 875–880. [CrossRef]
62. Roth, G.-L.; Esen, C.; Hellmann, R. Femtosecond Laser Direct Generation of 3D-Microfluidic Channels inside Bulk PMMA. *Opt. Express* **2017**, *25*, 18442–18450. [CrossRef] [PubMed]

63. Khorshidian, H.; James, L.A.; Butt, S.D. Demonstrating the Effect of Hydraulic Continuity of the Wetting Phase on the Performance of Pore Network Micromodels during Gas Assisted Gravity Drainage. *J. Pet. Sci. Eng.* **2018**, *165*, 375–387. [CrossRef]
64. Soleimani, M. Naturally Fractured Hydrocarbon Reservoir Simulation by Elastic Fracture Modeling. *Pet. Sci.* **2017**, *14*, 286–301. [CrossRef]
65. Liu, J.; Xie, H.; Wang, Q.; Chen, S.; Hu, Z. Influence of Pore Structure on Shale Gas Recovery with CO_2 Sequestration: Insight Into Molecular Mechanisms. *Energy Fuels* **2020**, *34*, 1240–1250. [CrossRef]
66. Auset, M.; Keller, A.A. Pore-Scale Processes That Control Dispersion of Colloids in Saturated Porous Media. *Water Resour. Res.* **2004**, *40*, W03503. [CrossRef]

Publisher's Note: MDPI stays neutral with regard to jurisdictional claims in published maps and institutional affiliations.

Article

Improvement of Etching Anisotropy in Fused Silica by Double-Pulse Fabrication

Valdemar Stankevič *, Jonas Karosas, Gediminas Račiukaitis and Paulius Gečys

Center for Physical Sciences and Technology, Savanoriu Ave 231, LT-02300 Vilnius, Lithuania; jonaskarosas919@gmail.com (J.K.); g.raciukaitis@ftmc.lt (G.R.); p.gecys@ftmc.lt (P.G.)

* Correspondence: valdemar.stankevic@ftmc.lt

Received: 20 April 2020; Accepted: 6 May 2020; Published: 8 May 2020

Abstract: Femtosecond laser-induced selective etching (FLISE) is a promising technology for fabrication of a wide range of optical, mechanical and microfluidic devices. Various etching conditions, together with significant process optimisations, have already been demonstrated. However, the FLISE technology still faces severe limitations for a wide range of applications due to limited processing speed and polarization-dependent etching. In this article, we report our novel results on the double-pulse processing approach on the improvement of chemical etching anisotropy and >30% faster processing speed in fused silica. The effects of pulse delay and pulse duration were investigated for further understanding of the relations between nanograting formation and etching. The internal sub-surface modifications were recorded with double cross-polarised pulses of a femtosecond laser, and a new nanograting morphology (grid-like) was demonstrated by precisely adjusting the processing parameters in a narrow processing window. It was suggested that this grid-like morphology impacts the etching anisotropy, which could be improved by varying the delay between two orthogonally polarized laser pulses.

Keywords: femtosecond; fused silica; double pulses; selective chemical etching

---

1. Introduction

Within the past two decades, femtosecond pulse processing of transparent materials has demonstrated outstanding results [1,2]. Many various processing technologies were developed for marking, dicing, cutting and internal modifications of transparent materials. Femtosecond laser-induced selective etching (FLISE) was demonstrated for the first time in 2001 by Marcinkevičius et al. when a track recorded by a femtosecond laser inside fused silica was selectively etched along the modified region [3]. That was the start of subtractive manufacturing in transparent materials, mainly in bulk fused silica by forming microchannels and complex 3D structures. It was found, within two years, that the modifications are composed of self-organised nanogratings oriented perpendicular to the laser polarisation [4]. The detailed investigations showed that selective etching is related to the nanograting orientation [5] and molecular oxygen in nanopores within the nanogratings [6]. The latter discoveries prompted many FLISE optimisation works to be conducted [7–10], and FLISE has become a very promising technology for a wide range of complex applications, such as internal 3D structures for mechanical, optical and microfluidic devices [11–13]. The most used and investigated material for laser-induced chemical etching is fused silica.

New opportunities came from double-pulse fabrication approaches recently introduced for the processing of various materials. Double pulses with different wavelengths [14] or double-pulse laser technology for material ablation [15] gained the increased attention for the processing of transparent materials [16]. Pengjun et al. demonstrated that the temporal shaping of the femtosecond pulses could provide higher etching selectivity, and, at some defined pulse energies, the etching rate could be a few times higher compared to conventional single pulses [17] due to the better efficiency of photon absorption. The promising results were also achieved by combining double pulses with a linear and circular polarisation where the improvement in the etching rate for the formation of high aspect ratio channels was demonstrated [18]. In most publications it is assumed that the first pulse is responsible for the nanograting orientation [14,19]. However, the real situation could be more complicated. The nanograting formation in fused silica is usually related to the enhanced birefringence. Atoosa et al. demonstrated that orthogonally-polarised double-pulses could reduce the birefringence because the nanograting orientation is determined by the writing pulse with a higher intensity [20]. More recent work also confirmed that the second pulse could rewrite the nanograting orientation when the double pulses with non-equal energies are applied [21]. However, there are still limitations of the FLISE technology due to etching anisotropy (etching rate dependence on laser writing direction at a constant linear polarisation) and low processing (laser writing and etching) speed. The efforts to overcome this drawback was already made for single-pulse processing by varying the polarization direction during fabrication [22], or by changing laser pulse duration to the picosecond range [23]; however, the problem was not solved completely.

In this work, we investigate the double-pulse processing approach with crossed polarisations and variable inter-pulse delay. Initially, peculiarities of nanograting morphologies recorded with double pulses were investigated. Hereafter, we show the influence of the pulse duration and inter-pulse delay to the selective etching of the microchannels. In most applications, the laser writing trajectory has a curved shape. The nanogratings are statically orientated along the curved patch. Therefore, at different trajectory position, the nanograting orientation is shifted relative to the trajectory vector, and, as a consequence, the orientation-dependent etching takes place. To overcome this drawback, the double pulses with crossed polarisations were used, and the etching anisotropy and etching rate improvement were demonstrated. Particular experiments were arranged to fabricate the vertical bow-like structures—two-dimensional structures formed by raster scanning of a bow-like horizontal shape starting beneath the sample bottom surface and moving laterally through the laser focus up to the top sample surface. To our best knowledge, we demonstrate the new grid-like nanograting morphology for the first time and the possible explanation provided. That is an entirely new phenomenon that was not described earlier.

2. Materials and Methods

The setup for double-pulse experiments is schematically illustrated in Figure 1. The micromachining workstation with an integrated femtosecond laser (Pharos-6W, Light Conversion, Vilnius, Lithuania) was used. The ultrashort laser operated at a 1030 nm wavelength, and the 500 kHz pulse repetition rate was mainly used during the experiments. The laser was equipped with a tunable compressor which allowed the pulse duration to be tuned in the range of ~0.26 to ~10 ps with a positive or negative chirp. The laser power was controlled using an external attenuator which was calibrated to vary the mean laser power linearly. The laser beam was split into two pulses with a polarising beam splitter (PBS). The energy ratio between two pulses was controlled by rotating a $\lambda/2$ phase plate. During the tests, the energy ratio was set to 1:1. The Mach–Zehnder interferometer setup, composed from a variable arm (VA) and reference arm (RA), was used to combine two laser beams with a controllable temporal delay between pulses. The polarisation in the reference arm was set to $\boldsymbol{E_y}$ and that in the variable arm to $\boldsymbol{E_x}$. Afterwards, the beam diameter was reduced to ~2 mm using a two-lens telescope system and focused to the sample using the 100× microscope objective (M Plan NIR, Mitutoyo, Kanagawa, Japan) with a numerical aperture (NA) of 0.5. The microscope objective was translated with a linear translation stage (ANT130-L, Aerotech, Pittsburgh, PA, USA) along the Z direction. The 20× camera vision objective (Olympus Plan Achromat, Tokyo, Japan) was mounted on the same stage. A sample was mounted on a two-axis gimbal mount (GM200, Thorlabs, Mölndal, Sweden) and translated by a high-resolution (>500 nm) XY linear stage (ANT130-XY, Aerotech, Pittsburgh, PA, USA) with a scan speed up to 5 mm/s. The translation stages were controlled via the controller (A3200, Aerotech, Pittsburgh, PA, USA).

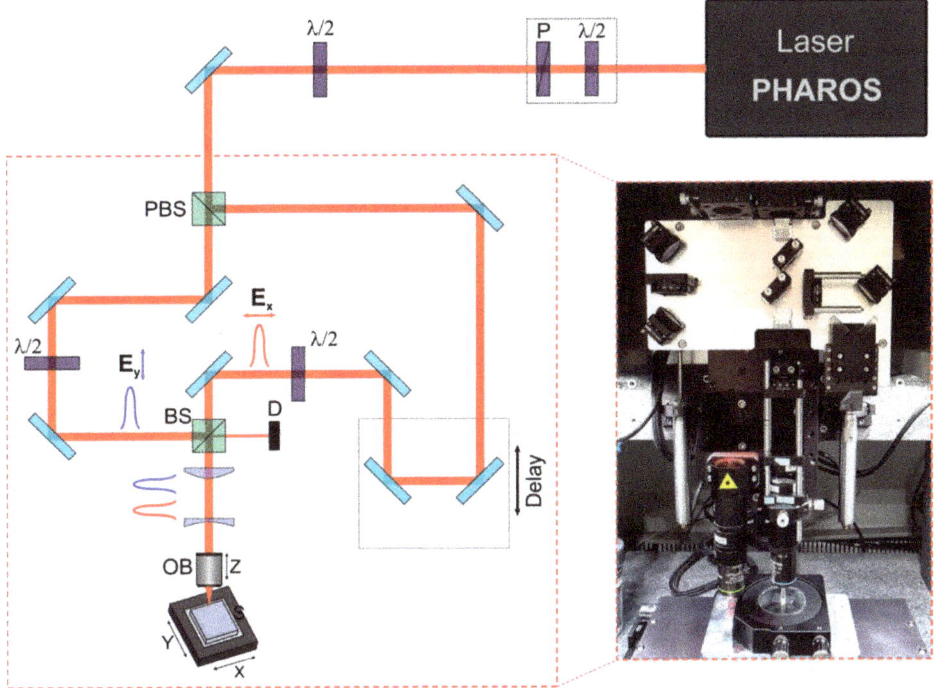

Figure 1. The double-pulse experimental setup. P—Brewster angle polarizer, PBS—polarising beam splitter, BS—beam splitter, D—dump, OB—microscope objective.

The delay range from –20 ps to 20 ps was possible to achieve, between the two pulses, a temporal delay resolution of ~7 fs. The delay was estimated relative to the reference arm; consequently, the negative delay was achieved when the variable arm was shorter with respect to the reference arm (the reference-arm pulse was the second). During the tests, the delay was varied from −10 ps to +10 ps. In further text, to avoid confusion, the following indication will be used: $E_{||}$: polarization parallel to the scanning direction, and $E_{\perp}$: polarization perpendicular to the scanning direction. The pulse order is indicated as the first or second one. To obtain the 0 fs delay, the double-pulse setup was calibrated temporally (Figure 2). Initially, the temporal delay was calibrated by changing the length of the delay line with a slight angular misalignment between the beams and registering the intensity profile with a CCD camera (Spiricon SP620U, Ophir, Jerusalem, Israel). When the delay in the variable arm approached the ~0 fs delay (i.e., two laser pulses overlap entirely in time), the clear interference intensity pattern was registered. When the laser pulses overlapped only partly (~160 fs), the interference was still observed. However, the contrast was very poor. For relatively long delays (>300 fs), the laser beams did not interfere. The interference condition also was not satisfied for cross-polarised beams, and the non-interfering double-pulse beam profile was observed even at the 0 fs pulse delay.

The spatial beam position adjustment was performed by recording the interference pattern at the 0 fs delay and measuring the interference period. According to the relation of the interference pattern and the beam angle, $d = \lambda/2sin\theta$, it was possible to calculate the angle between two beams. The interference period increased when the angle decreased. Therefore, two beams appeared almost parallel after a few iterations of mirror angle adjustment.

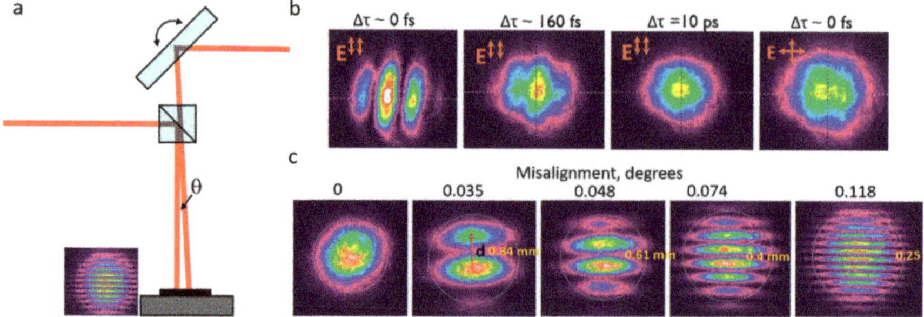

Figure 2. Spatial and temporal beam alignment: (**a**) two-beam angle adjustment principle; (**b**) intensity profiles of two beams with different temporal delay and polarisation orientation; (**c**) interference patterns of two beams with various misalignment angles at the 0 fs delay.

Commercially available fused silica (JGS1, 20 mm × 3 mm × 2 mm, Eksma Optics, Vilnius) samples were used for the experiments. The modifications were recorded by focused laser beam ~20 μm below the sample surface at the regimes when nanogratings were formed (Type II modification) [24]. A few parallel lines shifted by a z-step (1 μm) were recorded to ensure that the right vertical cross-section of the nanogratings would be investigated after the cutting and polishing procedure (Figure 3a). The exposed nanogratings were etched for 1 min in 5% hydrofluoric acid (HF) acid and coated with a ~10 nm gold layer. The prepared samples were investigated by scanning electron microscope (SEM) (JEOL, JSM6490LV, Tokyo, Japan).

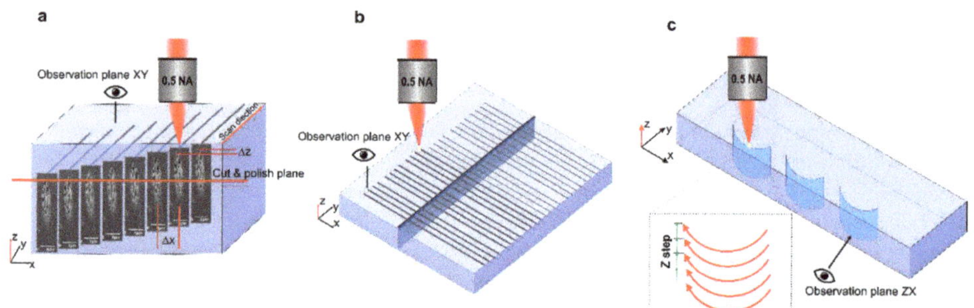

Figure 3. Schematic illustration of experiments: (**a**) recording and polishing of the lines written with double pulses for nanogratings observation by scanning electron microscope (SEM); (**b**) the line writing procedures for the investigation of the double-pulse influence on the chemical etching rate; (**c**) recording of the vertical bow-like structures for the investigation of directional etching dependence.

To investigate the double-pulse delay effects to the chemical etching rate of the fused silica samples, a group of lines was written by varying the pulse energy and scan speed at a constant repetition rate. The separate lines were recorded at different focusing depth (50–500 μm below the sample surface) with the vertical raster-scanned surface ending at the sample top surface in the middle of the recorded track to get the acid access to the modified area (Figure 3b). Three lines in each group were recorded under the same conditions and the measured data were averaged. The directional etching dependence on the double-pulse delay was investigated recording the bow-like vertical structures (Figure 3c) by a raster scanning the single bow-like (curved) trajectory laterally moving through the laser focus with a 7 μm vertical z step, starting beneath the bottom sample surface and ending on the sample top surface.

After the irradiation with the focused ultrashort laser pulses, the fused silica samples were immersed into a wet-etching bath of HF with a concentration of 10% and 30 °C temperature for 30 min.

After the chemical etching, the samples with the vertical bow-like structures were polished in the XZ plane with a 0.3 μm grade colloid silica to recover a smooth, transparent surface for the high-contrast microscope measurements.

3. Results and Discussion

3.1. Etching Rate of Modifications Recorded with Double Femtosecond Pulses

Multiple lines were inscribed in the bulk of fused silica, as shown in Figure 3a to investigate the effects of the inter-pulse delay, pulse energy, and focusing depth to the etching rate of the double-pulse modified material. The group of lines were recorded with different pulse density ranging from 100 to 2000 ppμ (pulses/μm), keeping the laser pulse energy constant. Various laser pulse energies were applied from 100 nJ to 600 nJ for each group of lines. It should be noted, that the indicated pulse energy was measured when two pulses were combined into one optical path, so the pulse energy for separate beams was twice lower. That was done to allow a direct comparison of the results with the single-pulse processing. In the single-pulse processing experiments, one of the beams was blocked.

In our previous works, the maximum etching rate of the fused silica was ~1300 μm/h with the leading etching selectivity of ~120:1, when processing with the 1030 nm wavelength was applied [25,26]. The results were achieved with ~400–600 nJ pulse energy and 1000 ppμ density. The results of the double-pulse processing in fused silica are demonstrated in Figure 4. The visual observation showed that even for double-pulse processing, the pulse duration had a strong influence on the etching rate. At 0 fs pulse delay and various pulse durations, we observed the etching rate variation when the pulse density was changed. A lower etching rate variation was observed for longer pulse durations independently on the chirp direction.

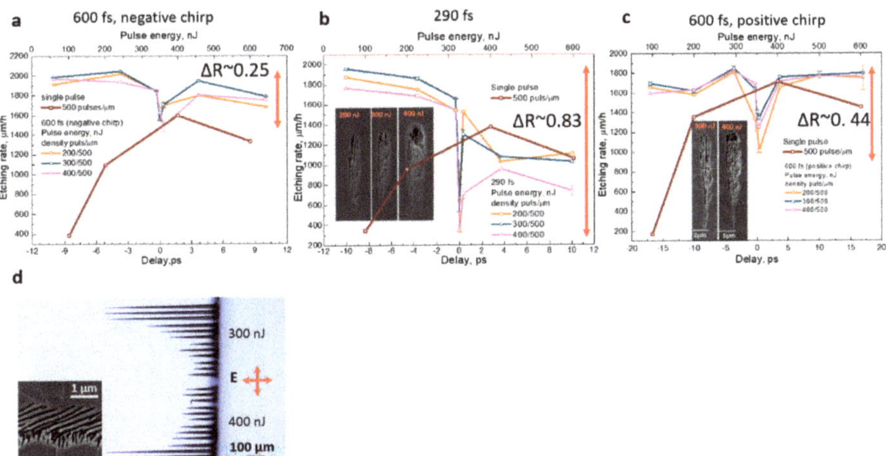

Figure 4. The etching rate dependence on the delay between two pulses for modifications recorded in fused silica with the 200–400 nJ pulse energy and orthogonal polarisations: (**a**) 600 fs pulse duration with a negative chirp; (**b**) 290 fs pulse duration; (**c**) 600 fs pulse duration with a positive chirp; (**d**) The microscope picture of etched channels for the 0 fs pulse delay at 290 fs pulse duration. Red curve: the etching rate dependence on the pulse energy for a single pulse. Samples were etched 30 min in 10% diluted hydrofluoric acid (HF) acid. The inset shows the SEM pictures from the side (**b**,**c**) and top (**d**).

The more surprising result was observed when the pulse delay was varied from −10 ps to +10 ps. As demonstrated in Figure 4a–c, the etching rate significantly reduced when the delay between two pulses approached 0 fs. For the positive and negative delays, the etching rate grew almost symmetrically when the pulse duration was 600 fs. For the shortest pulse duration (~290 fs with compensated chirp) and negative delays, the etching rate increase was higher compared to the positive delays. In the case of negative delay, the first beam coming from the variable arm was with the parallel polarisation ($E_{\| \text{first}}$), and only then the second beam from the reference arm with the perpendicular polarization ($E_{\perp \text{second}}$) was arriving. The etching rate difference could be related to the different modification thresholds for the nanograting formation, depending on the beam polarisation [6]. Therefore, the significant influence of the second pulse ($E_{\perp \text{second}}$) to the nanogratings orientation in the negative delay range was observed. Saturation of the etching rate was observed when the pulse delay approached ~10 ps independently on the delay direction. More objective etching rate comparisons could be made by introducing the etching rate contrast parameter ΔR, which is described as the ratio of the difference between the maximum and minimum etching rate and maximum etching rate as follows $\Delta R = (R_{max} - R_{min})/R_{max}$. According to this description, the etching contrast for the 600 fs pulse duration was consequently $\Delta R_{-600\text{fs}} = 0.25$ and $\Delta R_{+600\text{fs}} = 0.44$. The maximum etching rate contrast was achieved for the shortest pulse duration $\Delta R_{290\text{fs}} = 0.83$. The etching rate contrast showed how much the pulse delay influenced the etching rate: the lower etching rate contrast value, the weaker the pulse delay influence on the etching rate was observed. For longer pulse durations, the broader pulse delay range with a lower etching rate was found compared to the shorter pulse duration. Longer pulses had a wider temporal overlap range that confirms the observations of the highest etching rate drop only in the area where the temporal pulses overlap. It could be noted that the etching rate contrast drop was usually observed for higher pulse energies. By comparing the etching results of the single-pulse and double-pulse experiments, we can distinguish a significant difference in the etching rate. While for the single-pulse processing at the same pulse energy, the 1300–1400 µm/h etching rate was the maximum value that could be achieved, the double-pulse processing enhanced the etching rate up to 2000–2100 µm/h. That yielded >30% of the etching rate increase. This achievement is even more attractive as this rate enhancement can be achieved with a lower pulse density of 100–500 ppµ (pulse per micrometre), which allowed us to speed up the processing more than two times. Murata et al. demonstrated that for the double-pulse processing, the diameter of emerging nanopores is twice time larger comparing to the single-pulse processing [27]. We believe that such behaviour had substantial input to the etching rate increase in our case due to the higher area of nanopores.

To understand the etching rate enhancement, the in-depth investigations of the Type II modifications induced using double pulses at various delay times were performed. Figure 5 shows the nanograting morphology dependence on the delay between two pulses.

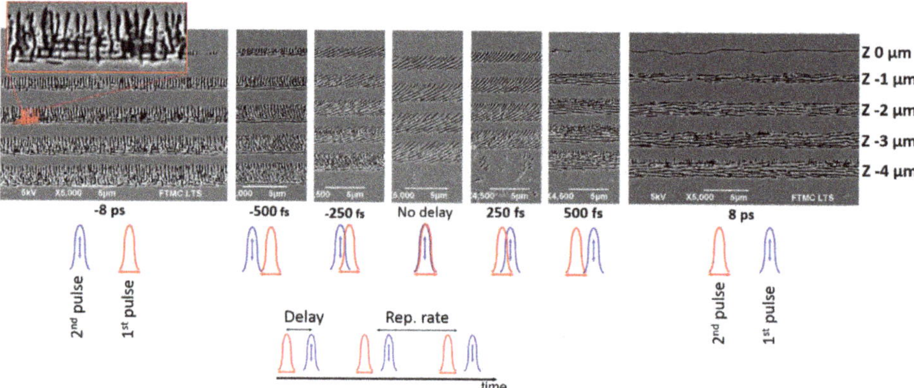

Figure 5. SEM pictures of the nanograting morphology evolution depending on the delay between two pulses. The inset shows the enlarged area of the grid-like nanograting structure. The nanostructures recorded with a total 400 nJ pulse energy (200 nJ + 200 nJ).

For the 0 fs delay between two pulses, the nanogratings consisted of parallel lines rotated by ~45° relative to the scanning direction. When temporal pulse overlap was in the range of laser pulse duration (±250 fs, for ~290 fs pulse duration), the 45° nanograting orientation was still observed. The nanogratings were slightly shifted from the straight line and appeared bow-shaped indicating the nonperfect spatial beam overlap. When the positive delay exceeded the pulse duration, the evident influence of the first pulse ($E_{\perp \text{ first}}$) on the nanograting formation was observed, that confirmed the results described by Rohloff et al. [28]. They found that, at low fluences, the orientation of the laser-induced periodic surface structures (LIPSS) structures on the fused silica surface changed their direction by 90° when delay direction changed, and it was determined by the first pulse polarization.

In our case, when the delay between laser pulses approached to −10 ps ($E_{|| \text{ first}}$), the grid-like structures were observed, which was an unexpected result considering the previous investigations [19,29]. This result was replicated a few times and was repeatable. To our best knowledge, the grid-like structures in the bulk fused silica were demonstrated for the first time, and there are no valuable phenomena to explain. We can speculate that the mentioned structures appear as a consequence of the different modification thresholds for two polarisations. According to previous investigations, the Type II modification threshold for perpendicular polarisation is ~2 times lower compared to the parallel polarisation [6]. The electron plasma generated by the first pulse absorbs more efficiently the second pulse due to the reduced ionisation threshold [30]. As a consequence, the already created point defects, such as colour centres (E′) and nonbridging oxygen hole centres (NBOHCs) [31], interact with the electrical field of the second pulse. Such interaction could enhance the electron plasma more effectively due to the lower modification threshold for perpendicular polarisation and a memory effect [30]. However, as it is known, at least a few tens of pulses are required to rewrite the nanogratings [32]. The polarisation changed after each pulse in the double-pulse regime, and that was not sufficient to overwrite the nanograting orientation. When the delay approached the −10 ps, the orientation of the inhomogeneities was created and defined by the electrical field orientation of the first pulse ($E_{|| \text{ first}}$) due to formed STE (self-trapped excitons) and not relaxed states still at a high temperature which involves the viscous flow of the silica. Hence, the induced periodic electron plasma pattern by the second pulse ($E_{\perp \text{ second}}$) with the perpendicular polarisation records the periodical nanopattern on top of the already oriented nanogratings partly covering them due to the lower modification threshold.

From another point of view, according to the numerical solution of Maxwell's equation in the vicinity of subsurface planar cracks oriented normal to the surface of fused silica [33] (created nanogratings in our case), the higher field enhancement is predicted (approx. by a factor of 2) inside narrow cracks (nanoplates) which are oriented perpendicular to the laser polarisation. In the case of negative delay ($E_{|| \text{ first}}$), the more significant field enhancement is predicted for the first pulse ($E_{|| \text{ first}}$) with the parallel polarisation (it first creates nanogratings perpendicular to the polarisation). However, due to a higher modification threshold for the same polarisation, the second pulse ($E_{\perp \text{ second}}$) should generate a higher amount of free electrons and create the grid-like nanostructures. In the case of positive delays ($E_{\perp \text{ first}}$), the influence of the first pulse strongly prevails due to higher light enhancement factor for the first pulse polarization (it first creates nanogratings perpendicular to the polarization) and lower modification threshold. Therefore, only a small part of the second pulse ($E_{|| \text{ second}}$) induced nanogratings is noticeable. However, for longer delays, the second pulse ($E_{|| \text{ second}}$) could have a stronger influence. As follows from Figure 4, the reduced etching rate for positive delays was observed.

The mechanism of the tilted nanogratings for the 0 fs delay can be simply explained in the following way: for spatially and temporarily adjusted beam the nanogratings are ~45° tilted relative to the X or Y polarisations, this is caused by the superposition of two pulses with orthogonal polarisations with no phase delay between two pulses, and the resultant vector is rotated by 45° [34,35]. The smallest step size of the delay line micrometre stage was 1 μm that changed the length of the delay line for double-length and corresponded to the ~7 fs delay. The resolution is not sufficient to precisely control the phase delay between two pulses, however for the exceptional case of 0 fs delay, the phase between two pulses could be set accurately due to the registration of the interference pattern of the slightly misaligned pulses. In this case, even the phase delay is set not accurately, and the elliptical polarization dominates, the morphology of the nanogratings still was normal to its major axis [36].

For the small beam misalignment in the +Y or −Y directions, the nanogratings were of the bow shape (Figure 6) due to partial interaction of the beams and domination of the opposite polarisations in the modification of top and bottom areas. For the intermediate case, the nanogratings were the result of both beams. For longer delays, the pulses did not interact coherently. Peculiarities in the nanograting morphology were defined by the interaction of the second pulse with the material excited by the first pulse.

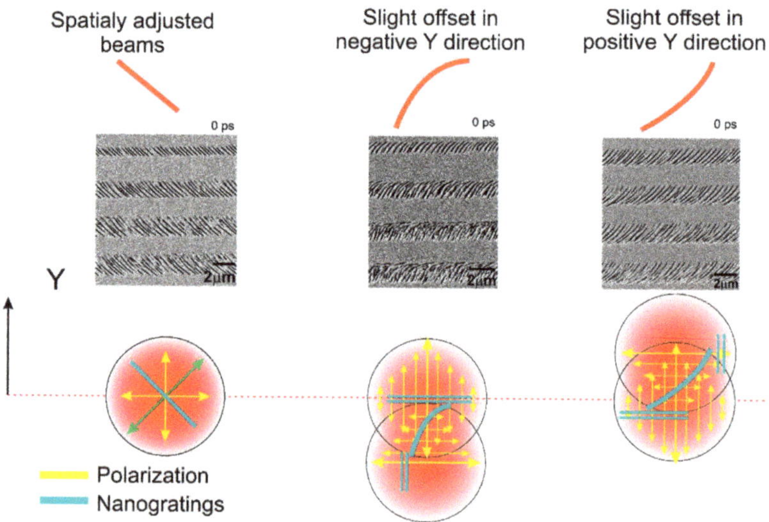

Figure 6. The pictorial explanation of the nanograting shape only for temporally adjusted pulses (0 fs delay) when a slight spatial beam misalignment is induced in the Y direction.

3.2. Directional Etching of Vertical Bow-Like Structures

In previous sections, we discussed the etching of the single microchannels recorded with the polarisation normal to the scan direction or utilising double-pulses with orthogonal polarisations. To investigate the double-pulse effect to the writing direction, bow-like vertical structures were inscribed in fused silica (Figure 3c). This way, the beam polarisation angle was constantly shifting relative to the writing direction of the bow-like curvature. Therefore, the different etching rate was obtained for the single-beam case depending on the location on the curve [22]. We were expecting that double pulses with orthogonal polarisations could suppress this effect due to grid-like structure formation and enhanced nanopores formation [27]. It should be considered, that in this recording configuration, additional influence of the line stacking along the beam propagation direction had an impact on the etching rate.

The vertical bow-like structures were recorded with the different double-pulse delay from −10 ps to +10 ps in fused silica. The separation between single vertical lines was set to 7 μm and was constant during all experiments. For one set of the double-pulse delay, an array of vertical surfaces with two different pulse densities of 500 and 1000 ppμ and four different pulse energies 200, 400, 600 and 800 nJ were recorded. The structures were etched 30 min in 10% HF acid. The etching behaviours of the vertical structures are demonstrated in Figure 7. Due to disturbance in the laser beam profile by the spherical aberration, a weak sample etching from the bottom surface was observed for the lower pulse energies (200–400 nJ). For higher pulse energies, the etching rate from both bottom and top surfaces was comparable at a delay in the range from −10 ps to 0 ps. However, the bottom structure etching rate was lower at the delay from 0 ps to 10 ps. The etching dependence on the direction was minimised for negative delays ($E_{\parallel\,\mathrm{first}}$) (for etching from the top and bottom surfaces). In the case of positive delays ($E_{\perp\,\mathrm{first}}$), the directional etching effect was significant, demonstrating the lower etching in the middle part of the vertical structure (Figure 8b, left inset).

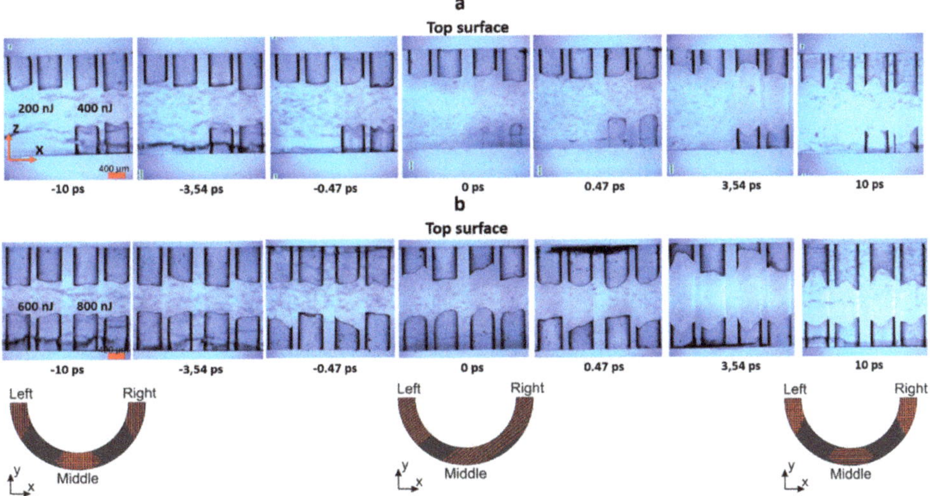

Figure 7. The microscope measurement of the etching dependence on the double-pulse delay for the bow-like vertical structures from the sample side (ZX plane): (**a**) the structures recorded with 200 nJ and 400 nJ pulse energy and (**b**) the structures inscribed with 600 nJ and 800 nJ pulse energy. Two channels with 500 and 1000 ppµ density were recorded at the constant pulse energy. Etching performed for 30 min. in 10% HF. The inset shows the bow-like trajectory view on the XY plane with a predicted nanogratings orientations.

For a more objective analysis, we introduce the etching isotropy factor K, which demonstrates the etching difference of the bow-like vertical structures in the middle part and corner parts (Figure 8a) $K = 2L_M/(L_R + L_L)$, where L_M, L_R and L_L are consequently the etched depth at the middle, right and left structure part. When K is approaching 1, uniform etching is obtained. For comparison, we recorded the bow-like vertical structures with a single pulse (Figure 8a). Usually, for this regime, the etching isotropy factor was below 0.7 and going down when the pulse energy or pulse density was decreased. For the 600 nJ pulse energy, the etching isotropy factors for single and double-pulse recording were 0.62 and 0.84, respectively. The 26% gain in the etching isotropy was obtained. The etching isotropy factor increased by raising the pulse energy for the positive pulse delays ($E_{\perp\,\mathrm{first}}$) (Figure 8b). However, $K > 1$ value was observed in the delay range from −0.3 ps to 0.3 ps, i.e., within the temporal pulse overlap range. The etching rate measurement also demonstrated that the maximal etching rate of ~ 1500 µm/h was achieved usually for negative delays ($E_{\parallel\,\mathrm{first}}$). The difference in etching rate was marginal and close to the maximal value for the positive delays ($E_{\perp\,\mathrm{first}}$) in some cases. In comparison for the single-pulse regime, the maximal achieved etching rate was ~1000 µm/h, which is ~33% slower and agrees well with the single channels etching results. Therefore, the nanograting orientations in the double-pulse regime played a significant role and enabled easier etchant penetration to the modified area. The etching behaviour for different delays could be confirmed by the nanograting investigation in Figure 5, where the grid-like nanogratings were demonstrated. On the left or right side of the bow-like structure, the speed vector made up the different angle with the polarisation. When on the middle part the first pulse polarisation is parallel ($E_{\parallel\,\mathrm{first}}$), on the left or right side it tends to be perpendicular, that made the nanogratings to be along the scan direction. That behaviour suppressed the directional etching dependence. Contrary, in the positive delays ($E_{\perp\ \mathrm{first}}$), the first pulse influence should be leading in the middle part of the surface, where the nanogratings along the scan vector should be formed, consequently at right and left side the grid-like nanogratings were pronounced. According to this description, the faster etching rate was predicted in the middle structure part, as we can observe for the pulse delays from 0 ps to +0.47 ps. However, for the longer positive delays, the opposite behaviour was found. We could speculate, that the second pulse ($E_{\|\ \mathrm{second}}$) also had a substantial impact on the nanograting formation, where the middle structure part was less etched. Especially the second pulse influence to the nanogratings direction was observed for the longer delays (>10 ps) and higher second pulse energy [21].

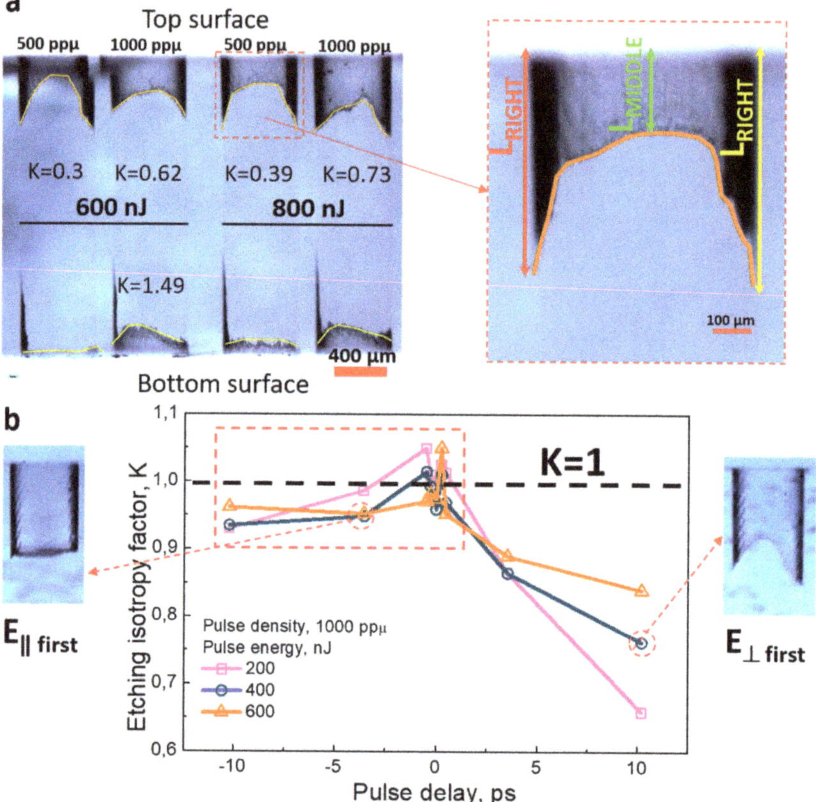

Figure 8. Etching of bow-like structures fabricated with the single-pulse regime at 600 nJ and 800 nJ pulse energy (**a**) and dependence of the etching isotropy factor K on the pulse delay for etching from the top sample surface of the vertical surface. (**b**). The insets show the microscope pictures with etched vertical surfaces.

4. Conclusions

The etching peculiarities of modifications in the fused silica recorded with the variable delay double-pulses were observed and discussed. The double-pulse fabrication enhanced the etching rate by >30% compared to the single-pulse fabrication. The detailed analysis of the nanostructures in the fused silica revealed the appearance of the grid-like nanostructures when the first pulse polarization was along scan direction ($E_{\|\ \mathrm{first}}$, negative delay). The nanogratings morphology explains the etching rate dependence on the pulse delay. The etching rate raised for negative and positive delays and was suppressed at the 0 fs delay, where the 45° oriented nanostructures are formed. The application of double pulses significantly improved the etching isotropy of the vertical bow-like structures. It was measured qualitatively by introducing the etching isotropy factor that showed the value near 1 for negative delays. We demonstrated that the double-pulse processing technique is a simple fabrication method that improves the etching isotropy without a need for phase plate rotation during processing. The etching rate improvement is even more attractive as this rate enhancement can be achieved with lower pulse density of 100–500 ppμ, which allows to speed up the processing more than two times.

Author Contributions: Conceptualization: V.S.; methodology: V.S. and J.K.; investigation: V.S. and J.K.; validation: V.S., P.G. and G.R.; formal analysis: V.S.; resources: V.S.; data curation: V.S.; writing—original draft preparation: V.S.; writing—review and editing: P.G. and G.R.; visualization: V.S and P.G.; supervision: P.G. and G.R. All authors have read and agreed to the published version of the manuscript.

Funding: This research received no external funding.

Conflicts of Interest: The authors declare no conflict of interest.

References

1. Shimotsuma, Y.; Hirao, K.; Qiu, J.; Miura, K. Nanofabrication in transparent materials with a femtosecond pulse laser. *J. Non-Cryst. Solids* **2006**, *352*, 646–656. [CrossRef]
2. Gattass, R.R.; Mazur, E. Femtosecond laser micromachining in transparent materials. *Nat. Photonics* **2008**, *2*, 219–225. [CrossRef]
3. Marcinkevičius, A.; Juodkazis, S. Femtosecond laser-assisted three-dimensional microfabrication in silica. *Opt. Lett.* **2001**, *26*, 277–279. [CrossRef]
4. Shimotsuma, Y.; Kazansky, P.G.; Qiu, J.; Hirao, K. Self-organized nanogratings in glass irradiated by ultrashort light pulses. *Phys. Rev. Lett.* **2003**, *91*, 247405. [CrossRef] [PubMed]
5. Hnatovsky, C.; Taylor, R.S.; Simova, E.; Bhardwaj, V.R.; Rayner, D.M.; Corkum, P.B. Polarization-selective etching in femtosecond laser-assisted microfluidic channel fabrication in fused silica. *Opt. Lett.* **2005**, *30*, 1867–1869. [CrossRef]
6. Lancry, M.; Poumellec, B.; Canning, J.; Cook, K.; Poulin, J.C.; Brisset, F. Ultrafast nanoporous silica formation driven by femtosecond laser irradiation. *Laser Photonics Rev.* **2013**, *7*, 953–962. [CrossRef]
7. Bellouard, Y.; Said, A.; Dugan, M.; Bado, P. Fabrication of high-aspect ratio, micro-fluidic channels and tunnels using femtosecond laser pulses and chemical etching. *Opt. Express* **2004**, *12*, 2120. [CrossRef] [PubMed]
8. Ho, S.; Herman, P.R.; Aitchison, J.S. Single- and multi-scan femtosecond laser writing for selective chemical etching of cross section patternable glass micro-channels. *Appl. Phys. A* **2012**, *106*, 5–13. [CrossRef]
9. Turco, S.L.; Osellame, R.; Ramponi, R.; Vishnubhatla, K.C. Hybrid chemical etching of femtosecond irradiated 3D structures in fused silica glass. *MATEC Web Conf.* **2013**, *8*, 05009. [CrossRef]
10. Hermans, M.; Gottmann, J.; Riedel, F. Selective, laser-induced etching of fused silica at high scan-speeds using KOH. *J. Laser Micro Nanoeng.* **2014**, *9*, 126–131. [CrossRef]
11. Bellouard, Y.; Said, A.A.; Bado, P. Integrating optics and micro-mechanics in a single substrate: A step toward monolithic integration in fused silica. *Opt. Express* **2005**, *13*, 6635–6644. [CrossRef] [PubMed]
12. Sugioka, K.; Xu, J.; Wu, D.; Hanada, Y.; Wang, Z. Femtosecond laser 3D micromachining: A powerful tool for the fabrication of microfluidic, optofluidic, and electrofluidic devices based on glass. *Lab Chip* **2014**, *14*, 3447–3458. [CrossRef] [PubMed]
13. Osellame, R.; Hoekstra, H.J.W.M.; Cerullo, G.; Pollnau, M. Femtosecond laser microstructuring: An enabling tool for optofluidic lab-on-chips. *Laser Photonics Rev.* **2011**, *5*, 442–463. [CrossRef]
14. Höhm, S.; Herzlieb, M.; Rosenfeld, A.; Krüger, J.; Bonse, J. Laser-induced periodic surface structures on fused silica upon cross-polarized two-color double-fs-pulse irradiation. *Appl. Surf. Sci.* **2015**, *336*, 39–42. [CrossRef]
15. Schille, J.; Schneider, L.; Kraft, S.; Hartwig, L.; Loeschner, U. Experimental study on double-pulse laser ablation of steel upon multiple parallel-polarized ultrashort-pulse irradiations. *Appl. Phys. A* **2016**, *122*, 644. [CrossRef]
16. Chu, W.; Tan, Y.; Wang, P.; Xu, J.; Li, W.; Qi, J.; Cheng, Y. Centimeter-Height 3D Printing with Femtosecond Laser Two-Photon Polymerization. *Adv. Mater. Technol.* **2018**, *3*, 1–6. [CrossRef]

17. Liu, P.; Jiang, L.; Hu, J.; Yan, X.; Xia, B.; Lu, Y. Etching rate enhancement by shaped femtosecond pulse train electron dynamics control for microchannels fabrication in fused silica glass. *Opt. Lett.* **2013**, *38*, 4613. [CrossRef]
18. Chu, D.; Sun, X.; Dong, X.; Yin, K.; Luo, Z.; Chen, G.; Duan, J.-A.; Hu, Y. Effect of double-pulse-laser polarization and time delay on laser-assisted etching of fused silica. *J. Phys. D Appl. Phys.* **2017**, *15*, 013001. [CrossRef]
19. Wang, H.; Song, J.; Li, Q.; Zeng, X.; Dai, Y. Formation of nanograting in fused silica by temporally delayed femtosecond double-pulse irradiation. *J. Phys. D Appl. Phys.* **2018**, *51*, 155101. [CrossRef]
20. Sadat, A.; Somayeh, A.; Aliasghar, N.; Wolfgang, A.; Reza, H. Birefringence profile adjustment by spatial overlap of nanogratings induced by ultra-short laser pulses inside fused silica. *Appl. Phys. A* **2018**, *124*, 1–6.
21. Zhang, W.; Zhai, Q.; Song, J.; Lou, K.; Li, Y.; Ou, Z.; Zhao, Q.; Dai, Y. Manipulation of self-organized nanograting for erasing and rewriting by ultrashort double-pulse sequences irradiation in fused silica. *J. Phys. D Appl. Phys.* **2020**, *53*, 165106. [CrossRef]
22. Ross, C.A.; MacLachlan, D.G.; Choudhury, D.; Thomson, R.R. Optimisation of ultrafast laser assisted etching in fused silica. *Opt. Express* **2018**, *26*, 24343. [CrossRef] [PubMed]
23. Li, X.; Xu, J.; Lin, Z.; Qi, J.; Wang, P.; Chu, W.; Fang, Z.; Wang, Z.; Chai, Z.; Cheng, Y. Polarization-insensitive space-selective etching in fused silica induced by picosecond laser irradiation. *Appl. Surf. Sci.* **2019**, *485*, 188–193. [CrossRef]
24. Hnatovsky, C.; Taylor, R.S.; Rajeev, P.P.; Simova, E.; Bhardwaj, V.R.; Rayner, D.M.; Corkum, P.B. Pulse duration dependence of femtosecond-laser-fabricated nanogratings in fused silica. *Appl. Phys. Lett.* **2005**, *87*, 014104. [CrossRef]
25. Stankevič, V. *Formation and Characterization of Micro-Opto-Mechanical 3D Devices for Sensor Application in Transparent Materials*; Vilnius University; Center for Physical Sciences and Technology: Vilnius, Lithuania, 2017.
26. Stankevič, V.; Rakickas, T.; Račiukaitis, G. Internal to External Microfluidic Device for Ellipsometric Biosensor Application. *J. Laser Micro Nanoeng.* **2016**, *11*, 53–58. [CrossRef]
27. Murata, A.; Shimotsuma, Y.; Sakakura, M.; Miura, K. Control of periodic nanostructure embedded in SiO_2 glass under femtosecond double-pulse irradiation. *J. Laser Micro Nanoeng.* **2016**, *11*, 95–99. [CrossRef]
28. Rohloff, M.; Das, S.K.; Höhm, S.; Grunwald, R.; Rosenfeld, A.; Krüger, J.; Bonse, J. Formation of laser-induced periodic surface structures on fused silica upon multiple cross-polarized double-femtosecond-laser-pulse irradiation sequences. *J. Appl. Phys.* **2011**, *110*, 014910. [CrossRef]
29. Höhm, S.; Herzlieb, M.; Rosenfeld, A.; Krüger, J.; Bonse, J. Dynamics of the formation of laser-induced periodic surface structures (LIPSS) upon femtosecond two-color double-pulse irradiation of metals, semiconductors, and dielectrics. *Appl. Surf. Sci.* **2016**, *374*, 331–338. [CrossRef]
30. Rajeev, P.P.; Gertsvolf, M.; Simova, E.; Hnatovsky, C.; Taylor, R.S.; Bhardwaj, V.R.; Rayner, D.M.; Corkum, P.B. Memory in Nonlinear Ionization of Transparent Solids. *Phys. Rev. Lett.* **2006**, *97*, 253001. [CrossRef]
31. Richter, S.; Jia, F.; Heinrich, M.; Döring, S.; Peschel, U.; Tünnermann, A.; Nolte, S. The role of self-trapped excitons and defects in the formation of nanogratings in fused silica. *Opt. Lett.* **2012**, *37*, 482–484. [CrossRef]
32. Taylor, R.; Hnatovsky, C.; Simova, E. Applications of femtosecond laser induced self-organized planar nanocracks inside fused silica glass. *Laser Photonics Rev.* **2008**, *2*, 26–46. [CrossRef]
33. Génin, F.Y.; Salleo, A.; Pistor, T.V.; Chase, L.L. Role of light intensification by cracks in optical breakdown on surfaces. *J. Opt. Soc. Am. A* **2001**, *18*, 2607. [CrossRef] [PubMed]
34. Hecht, E. *Optics*; Addison-Wesley: Boston, MA, USA, 2002; ISBN 0321188780.
35. *Springer Handbook of Lasers and Optics*; Springer: Berlin, Germany, 2007; Volume 45, ISBN 9780387955797.
36. Taylor, R.S.; Simova, E.; Hnatovsky, C. Creation of chiral structures inside fused silica glass. *Opt. Lett.* **2008**, *33*, 1312. [CrossRef] [PubMed]

Article

One-Step Femtosecond Laser Stealth Dicing of Quartz

Caterina Gaudiuso [1,*], Annalisa Volpe [2] and Antonio Ancona [1]

1 CNR-IFN UOS BARI, Via Amendola 173, 70126 Bari, Italy; antonio.ancona@uniba.it

2 Dipartimento Interateneo di Fisica, Università degli Studi di Bari, 70125 Bari, Italy; annalisa.volpe@uniba.it

* Correspondence: caterina.gaudiuso@uniba.it; Tel.: +39-0805442386

Received: 24 January 2020; Accepted: 21 March 2020; Published: 22 March 2020

Abstract: We report on a one-step method for cutting 250-µm-thick quartz plates using highly focused ultrashort laser pulses with a duration of 200 fs and a wavelength of 1030 nm. We show that the repetition rate, the scan speed, the pulse overlap and the pulse energy directly influence the cutting process and quality. Therefore, a suitable choice of these parameters was necessary to get single-pass stealth dicing with neat and flat cut edges. The mechanism behind the stealth dicing process was ascribed to tensile stresses generated by the relaxation of the compressive stresses originated in the laser beam focal volume during irradiation in the bulk material. Such stresses produced micro-fractures whose controlled propagation along the laser beam path led to cutting of the samples.

Keywords: ultrashort laser pulses; heat accumulation; transparent materials; quartz; stealth dicing

---

1. Introduction

Transparent materials are used in an increasing number of applications, ranging from microelectronics [1], to microfluidics [2,3] and optoelectronics [4]. In particular, glass, quartz and sapphire, due to their broad spectral band of light transmission, hardness and scratch-resistance, are good candidates for display components of portable mobile electronics, as light emitting diode (LED) substrates, protection mirrors of mobile phone cameras, smartwatches, etc. Despite their growing use in consumer electronics products, cutting these materials is still a challenging task, due to their brittleness. Obtaining a high-quality cut edge completely free from micro-cracks or chips is not easy, particularly when employing thinner substrates. The state of the art cutting technologies most widely used include traditional mechanical cutting, chemical etching, electrochemical machining and laser-based methods.

Traditional machining methods are mainly based on diamond cutting [4]. Here, the material is first marked and scribed with a diamond tool and then an external force is applied to break the substrate along the scribing path [5]. Unfortunately, with this method, at high cutting speeds the diamond blades can generate chipping and cracks, which compromise the quality of the cutting edges, reducing the resistance of the materials by up to 60% in the case of glass [6]. Moreover, tool wear affects the repeatability and the efficiency of the dicing process [7], and additional grinding and polishing steps are needed to achieve the required smooth finish.

Another method for cutting transparent materials is based on the selective chemical etching of laser modifications inside the volume [8]. The main advantage of this technique, besides the lack of debris, is the possibility of designing complex shapes. However, several modifications are required along the entire thickness of the substrate, thus limiting the processing speed. Furthermore, hydrofluoric acid (HF) is the only etchant which attacks amorphous SiO_2, quartz or glasses at significantly high etching rates [9]. Unfortunately, due to its high toxicity and corrosiveness, when hydrogen fluoride (HF) is used for large scale productions, many countries require very strict safety regulations, see e.g., [10,11].

Electrochemical methods for machining quartz were reported by several authors [12–15]. Jain et al. [12] exploited electrochemical spark machining (ECSM) showing that, depending on the polarity and the applied voltage, a cut kerf ranging approximately from 0.5 to 0.9 mm and a surface roughness from 3 to 14 μm were obtained on a 2 mm-thick quartz sample. Wang et al. [15] demonstrated the shape cutting of quartz glass by wire electrochemical discharge machining (WECDM), but with this technique, significant bulges were generated on the edge of the cut circle.

In laser processing methods, the energy of a focused laser beam is exploited to modify the substrate from its surface or in the bulk, thus leading to its separation. Laser-based techniques are making huge advances in the field of cutting transparent materials, as they have numerous advantages over mechanical methods. Being non-contact, laser processes avoid any effect due to tool wear and mechanical stress, achieving high quality and precision of the cut at reasonable costs. Furthermore, the use of the laser prevents any contamination of the materials being processed. However, the process parameters, and in particular the wavelength of the laser source and its pulse width, must be carefully selected and adjusted to obtain the desired results and cutting-edge quality [16].

One of the first laser-based methods was introduced by Garibotti in 1963 [17] and consists in laser scribing and dicing. It is a two-step process: the laser is focused on the surface of the workpiece to generate a groove, which is subsequently fractured by applying a tensile stress. The mechanical stress that initiates the cutting process can be generated either by an external mechanical force or by a rapid heating-cooling cycle [10]. Serdyukov et al. [18] reported on numerical simulation about the controlled thermal cleavage of crystalline quartz, produced by laser heating and exposure to coolant. The cut was ascribed to the thermoelastic stress produced in the quartz crystals, due to the anisotropy of the thermal conductivity and thermal expansion. Such interpretation was confirmed through experiments performed using a CO_2 laser, which led to the successful cut of quartz crystal, though with the presence of evident chips and micro-cracks.

In some cases, the laser irradiation itself provides enough thermal stress to cause the cutting of the sample. Such cutting process is referred to as thermal cleavage [19]. The focused laser spot acts, indeed, as a localized heat source generating a thermo-mechanical compressive stress, whose relaxation causes the material to separate along the laser scanning path [20]. The fracture mechanism is similar to a crack extension, which can deviate from the desired path, especially for fast cutting speeds and long cutting lengths [11]. After being optimized in order to control the path of the stress-induced fracture, such a technique has been successfully applied to cut various materials, including silicon [21], alumina ceramics [22] and glass [23]. Xu et al. [19] demonstrated the laser thermal cleavage of sapphire substrate wafers using a CO_2 laser source. Here, a groove was engraved by laser ablation on the substrate surface. Laser irradiation caused localized and rapid heating and cooling, which led to the generation of local micro-cracks that propagated along preferential directions, with consequent cleaving and cutting of the substrate. Although attractive, this experimental procedure is very complex, since it requires cooling of the substrate and accurate alignment of the laser engrave to achieve the desired cutting path. Moreover, a protective layer must be applied to avoid surface contamination by redeposited laser ablation debris. An indirect, two steps thermal cleavage process has been developed by Choi et al. [16] for cutting glass, using a near-infrared (NIR) nanosecond laser. In this case, a laser-induced plasma plume was produced by irradiating a sacrificial absorbing layer, positioned a few hundred microns far from the glass target. Such a plume allowed the improvement of the absorption of laser energy by the target substrate, thus achieving localized and rapid heating and cooling. By delivering the appropriate amount of laser energy generating the plasma plume, controlled local microcracks were induced, which resulted in the cutting of the glass target.

A different approach was used by Russ et al. for cutting thin and ultra-thin glass plates (i.e., a thickness of a few hundred micrometers). Using ultrashort laser pulses, they ablated, layer-by-layer, the entire thickness of the substrate along the desired contour path [24]. This approach does not require applying any tensile stress and is suitable for any geometry, however it does not produce perpendicular cutting edges, has limited processing speed and produces a large amount of ablation debris [25]. Vanagas et al. used an analogous approach for the cutting of quartz and borosilicate cover glass samples, using 150 fs laser pulses at the wavelength of 800 m and repetition rate of 1 kHz.

They demonstrated the feasibility of the laser cutting of quartz, although the processing speed was 200 μm/s and the laser-induced damage on the rear surface, caused by the multiple overlapped scanned paths, was quite extended [26].

Stealth dicing, instead, consists of focusing the laser beam inside the bulk material, transparent to the laser wavelength, and moving it along the desired path that acts as the initial division line, when an external tensile stress is subsequently applied. The process is ablation-free, does not generate any debris, and is extremely fast. The successful stealth dicing of thin sapphire wafers (350 μm) has been demonstrated using a fs-laser in the NIR wavelengths focused through a microscope objective [1]. However, single-focus stealth dicing laser processing does not always ensure a precise control of the taper, thus compromising the quality of the cutting edge. The laser cutting of thin borosilicate glass slides using a commercial fs-laser has been tested [27], finding that the fine cut can be successfully obtained by carefully adjusting the scan speed, in order to trigger the formation of micro-cracks at the exit of the laser beam from the sample. The generation of such microcracks was ascribed to the damage induced by filamentation and the consequent mechanical stress built-up in the material. Moreover, many studies have explored multifocal laser processing, mainly using a combination of diffractive optical elements (i.e., Fresnel lens) and Bessel beams [22,23]. Tsai et al. [28] demonstrated the cutting of a thin glass with a thickness of 100 μm through modification in the bulk volume, using a femtosecond laser Bessel beam and applying a breaking stress. A smooth cut edge with chipping <1 μm was obtained. Dudutis et al. [29] proved the possibility to use a picosecond laser for glass dicing using an axicon-generated asymmetrical Bessel beam. A tilt movement was applied to the axicon holder in order to add an astigmatic aberration, producing a Bessel beam with controlled asymmetry. This solution allowed the achievement of 2.8 times faster dicing speed and the lower propagation of cracks into the bulk material, compared to dicing with a symmetrical Bessel beam. The stealth dicing of sapphire using ultrashort pulsed Bessel beams has also been demonstrated by Lopez et al. [30], but achieving a sidewall roughness of around 2 μm.

In this work, we investigate the cutting of quartz by ultrashort laser pulses. Quartz is a material of relevant interest in many fields like e.g., optoelectronics, fibre-technology and photoacoustics [31,32], thanks to its high optical transmission in a broad range from ultraviolet (UV) to mid- infrared (MIR), unique thermal and electrical properties, besides an excellent chemical resistance. The high precision cutting of quartz is crucial for many applications that require the fabrication of miniaturized quartz-based devices. The traditional method for cutting quartz is based on lithography followed by wet etching, using highly toxic agents like ammonium bifluoride [17] or HF acid [8]. The laser-based cutting method proposed in this work is a clean, single pass stealth dicing process, which does not involve any chemical agent or external tensile stress. The dependence of the main laser process parameters as the repetition rate, the pulse overlap and the pulse energy on the cutting efficiency were experimentally investigated. The quality of the cut edge has been thoroughly analyzed by optical microscopy. Optical profilometry has been exploited to evaluate the roughness of the cut edge and the flatness of the final cuts.

2. Materials and Methods

The set-up used for these experiments is shown in Figure 1. The laser source was the Pharos SP 1.5 from Light Conversion, providing 200 fs pulses with variable repetition rate from a single pulse to 1 MHz. The almost diffraction limited laser beam (M^2<1.3) was characterized by a central wavelength of 1030 nm and had a maximum average power of 6 W and maximum pulse energy E_p of 1.5 mJ.

The linearly polarized exit beam passed through a half-wave plate and a polarizer, which allowed one to tune the average power by rotating the half waveplate. Next, the laser beam was sent to a microscope objective with a focal length of 8 mm (the estimated focused spot diameter d in air was 1.3 μm) mounted on a PC controlled motorized axis (Aerotech, ANT130 LZS, Pittsburgh, PA, USA). This enabled the beam focus to be finely positioned in the bulk of the transparent samples that were moved on a XY plane, perpendicular to the beam axis by two Aerotech ABL1500 motorized stages with sub-micrometer resolution. As samples, 250 μm thick Z-cut quartz plates from Nano Quartz Wafer were used. In the present experimental conditions, self-focusing is likely experienced at a distance from the sample surface estimable as [33]:

$$z_{sf} = \frac{2n_0(d/2)^2}{\lambda}\frac{1}{\sqrt{P/P_{cr}}} \tag{1}$$

where n_0 is the linear refractive index and λ the wavelength, P is the applied peak power and P_{cr}=11 MW is the critical power for having self-focusing, determined according to Eq. 7.1.1 in [33]. Considering that the applied peak power ranged from 75 MW to 175 MW, the calculated self-focusing distance ranged between 0.31 μm and 0.47 μm. Correspondingly, the spot radius at such distance was determined according to [34]:

$$w_{sf}^2(z_{sf}) = (d/2)^2\left[\left(1+\frac{z_{sf}}{R_0}\right)^2 + \frac{4\gamma}{k^2(d/2)^4}z_{sf}^2\right] \tag{2}$$

where R_0 is the curvature radius, k is the wave number in linear media and $\gamma = \left(1+\alpha^{-2}\right)\left(1-\frac{P}{P_{cr}}\right)$, with α the degree of global coherence. In the present case, considering $\alpha\rightarrow\infty$, the estimated beam spot radius due to self-focusing varied from 0.42 μm to 0.52 μm at the self-focusing distances, according to the applied peak power. However, it is important to highlight that the aberration occurring from focusing the laser beam deep inside the quartz sample could cause deviations from a diffraction-limited beam spot, by extending the focal region along the optical axis.

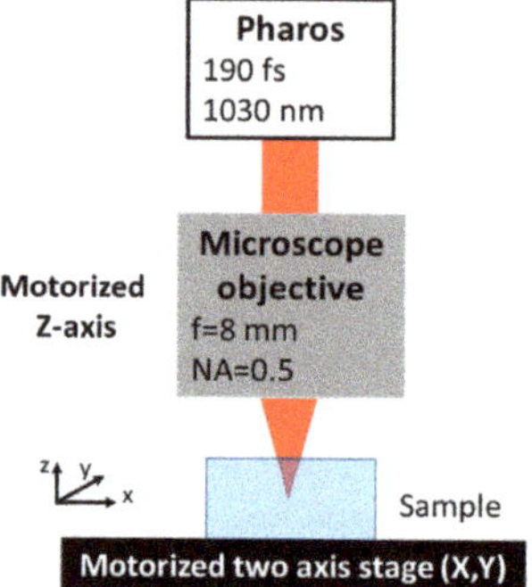

Figure 1. Experimental set-up used for the cutting experiments.

The translation speed v and the repetition rate f defined the average number of pulses pps (pulses per spot) impinging on the same focal area, calculated as pps = w_{sf}*f/v.

The repetition rate, the scan speed and the pulse energy were varied and their influence on the cutting process was evaluated, in order to better understand the underlying physical mechanisms.

3. Results and Discussion

3.1. *Influence of the Repetition Rate and pps*

It was found that the repetition rate plays a fundamental role to get the laser-induced dicing of the quartz plates in a single step. Indeed, the repetition rate defines the time separation between two successive laser pulses and together with the travel speed and the pulse energy, determines the amount of laser energy and heat released in a given material volume per unit time.

Using a repetition rate of 25 kHz, the single-step cutting of quartz was never achieved regardless the travel speed and, thus, the number of pps. At higher repetition rates, stealth dicing of the plates occurred with quite different cutting edge qualities, depending on the pulse energy and the number of pps.

In Figure 2, the optical microscope images of the cut edges obtained on quartz at 50 kHz and 100 kHz, with a pulse energy of 20 μJ and pps of 48 and 96, respectively, are shown. All the cuts exhibit a damaged area beneath the surface, in the region where the laser focus was placed. The areas above and below appear to be rather clean.

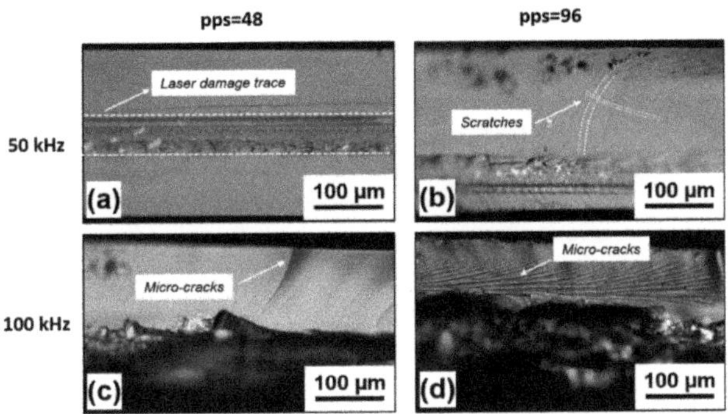

Figure 2. Optical microscope images of the cut edges obtained with a fixed pulse energy E_p=20 μJ and different repetition rates of 50 kHz and 100 kHz and pulse overlap pps= 48 and 96 (**a**–**d**).

Several physical mechanisms, generally taking place when irradiating a transparent material with intense ultrashort laser pulses, might have concurred to get the single-step stealth dicing of quartz. Besides self-focusing, also aberration from focusing deep inside the bulk quartz and filamentation typically occur at the peak power levels used in our experimental conditions [35]. However, the laser damaged area is quite confined inside the bulk material. Therefore, beam filamentation unlikely occurred along the whole quartz thickness. It is much more plausible that the physical mechanism originating the single pass stealth dicing is an accumulation of laser-induced stresses. In fact, in agreement with [36], it can be estimated that the peak temperature reached by the quartz lattice after each laser pulse is around 10^3 K. At such a high temperature, a transient tensile stress is generated, whose magnitude can be estimated by the following formula [16]:

$$\sigma = \frac{E\alpha\Delta\mathrm{T}}{1-\upsilon} \tag{3}$$

where E is the Young modulus, α is the coefficient of thermal expansion, ΔT is the temperature increase due to irradiation, and ν is the Poisson's number. The magnitude of such stress, which has been calculated taking into account the specific experimental conditions, is of the order of some hundreds of MPa. Literature reports a formation of cracks when the stress exceeds 1 GPa [37]. Therefore, the generation of a crack is not expected after each single pulse. However, since the single-pulse laser-generated mechanical stress lasts longer than 10 μs [38], if a second pulse arrives before such stress is released, then the two stresses accumulate. Pulse after pulse, this stress accumulation mechanism leads to the generation of cracks. The joining and propagation of cracks following the laser path and throughout the entire thickness of the quartz plate originates the single step stealth dicing process.

This explanation justifies the different results obtained at the three investigated values of repetition rate. In fact, at 25 kHz, the time delay between subsequent pulses is much longer than the relaxation time of each laser induced stress. Therefore, the stresses of following pulses do not overlap, and dicing of the sample does not take place. As far as the repetition rate increases, the pulse-to-pulse time interval shortens and the laser-induced stresses start to accumulate, finally resulting in the formation of cracks and dicing. At a repetition rate of 50 kHz, the time delay between consecutive pulses is half of the previous case and the self-induced stealth dicing of the samples was successfully obtained, with overall acceptable quality of the cut edges, especially for pps = 48.

In this case, the area modified by the laser interaction is well confined in the bulk volume and can be easily recognized, as shown in Figure 2a. Here, the average areal roughness Sa was found to be around 1 μm. No significant collateral damage is noticed above or below the laser modification trace and a much lower surface roughness of around 0.05 μm was found. By increasing the number of pps to 96, the quality of the cut edge is still acceptable, but some scratches begin to be noticed above and below the laser modified volume, see Figure 2b.

At an even higher repetition rate of 100 kHz, the laser induced crack propagation mechanism is no longer under control, owing to the excessive thermal load released into the focal volume [39]. This causes significant collateral damage around the laser absorption area, with unacceptable quality of the cut edges, as shown in Figure 2c,d. In particular, in the top part of the cut edge, long cracks propagating all the way towards the surface can be noticed at the lower number of pps = 48. For higher pps = 96, a large number of very dense erosion lines, accompanied by microcracks bridging them, appear in the upper part of the cut edge. For both investigated pps values, at 100 kHz of repetition rate, the lower part of the cut edge is completely destroyed, with big chips of quartz that have detached from the bottom.

The role of the overlap between pulses has been further investigated by performing additional experiments at the repetition rate of 50 kHz, keeping the same pulse energy of 20 μJ and reducing the number of pps to 24 and 10, respectively. Even in these two cases, the single step laser stealth dicing of the quartz samples was achieved. The corresponding optical microscope images of the cut edges are shown in Figure 3. As for pps = 48, the average areal roughness Sa of the laser modified zone was equal to 1 μm for both samples, while the surrounding area showed a surface roughness around 0.05 μm. However, compared to the case of pps = 48 shown in Figure 2a, where the cut edge was almost perfectly flat without any evident imperfection, a reduction of the number of pps has not led to further improvement of the cut quality. Unexpectedly, some small scratches or micro-cracks appear close to the laser modified area, suggesting that besides the repetition rate, the number of pps must also be selected within an ideal range to obtain a clean single step cut. On the other hand, a slight decrease of the laser damaged area depth, from approximately 50 μm to 35 μm, is noticed when decreasing the pps from 48 to 10. This is ascribable to the well-known incubation effect [40]. As the number of impinging pulses increases, the damage threshold decreases through the creation of point defects. Those defects enhance absorption of subsequent pulses, thus improving the coupling of laser energy into the lattice [40]. As a consequence, the dimensions of the laser-modified area increase with the number of pps. In addition, an increase of the pps leads to an increment of the accumulated fluence (defined as the pps multiplied by the fluence of the single pulse), which may change the morphology of the laser damage trace [41].

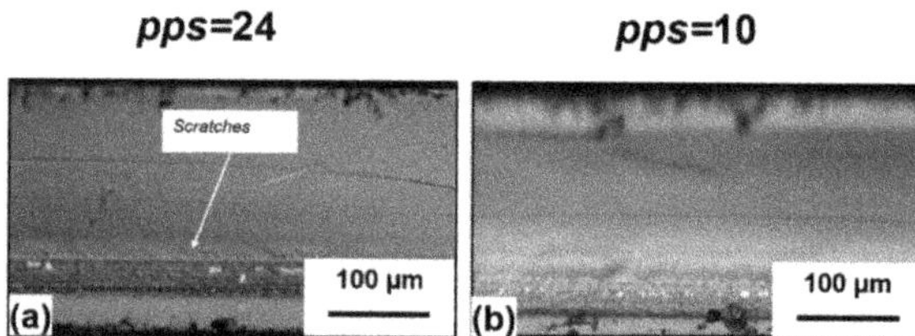

Figure 3. Optical microscope images of the cut edges obtained with pps = 24 and 10, at 50 kHz repetition rate and a pulse energy of 20 μJ. The reduction of the pps from 24 to 10 causes the cut edge not to be perpendicular to the target surface. In (**a**), the image is almost entirely on focus. In (**b**), the part of the cutting edge above the laser damage trace is out of focus, thus indicating a different height with respect to the part below.

3.2. *Influence of the Pulse Energy*

The influence of the pulse energy on the overall cutting mechanism and the cut edge quality has been investigated by carrying out experiments at 50 kHz of repetition rate, 1 mm/s of translation speed and varying the pulse energy from 15 to 40 µJ, with increments of 5 µJ. Indeed, below 15 µJ, the pulse energy was too low to get the single step cutting of the samples, thus indicating that it is a threshold process.

The corresponding cut edges are shown in Figure 4, where the top and bottom views are also presented. For Ep = 15 µJ (Figure 4a) a single step stealth dicing cut with acceptable quality of the cut edge was obtained, except for some small imperfections on the top of the laser modified area, which had a depth of 45 µm. The best cut quality was obtained at 20 µJ, where a clean cleavage is observed, with a 50-µm-deep laser modified zone buried inside the sample thickness. The cut is straight with well-defined edges following the laser path, and very few surface defects, as can be noticed from the top and bottom view of the sample, unlike the case reported by Vanagas et al. [26], where a damage area on the sample rear surface of approximately 500 µm wide was observed. A comparison with the top and bottom view of the same sample before irradiation (Figure 5) does not show any significant difference. Therefore, it can be excluded that when focusing the laser beam inside the bulk material surface, defects are generated on the top and the bottom of the sample.

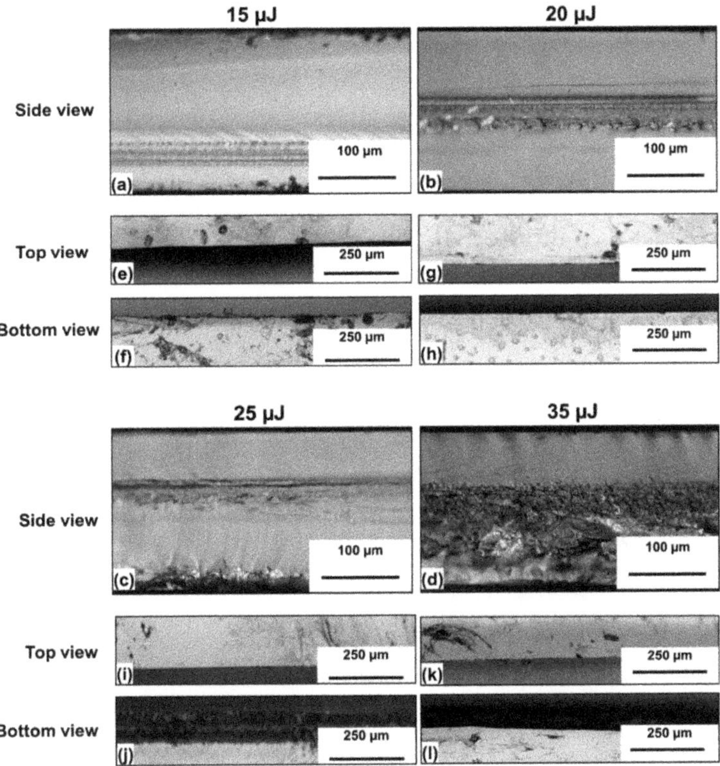

Figure 4. (**a–d**) Side view of the cut edges obtained at 50 kHz, at the scan speed of 1 mm/s and four different pulse energies. The top (**e**,**g**,**i**,**k**) and the bottom (**f**,**h**,**j**,**l**) views are also shown.

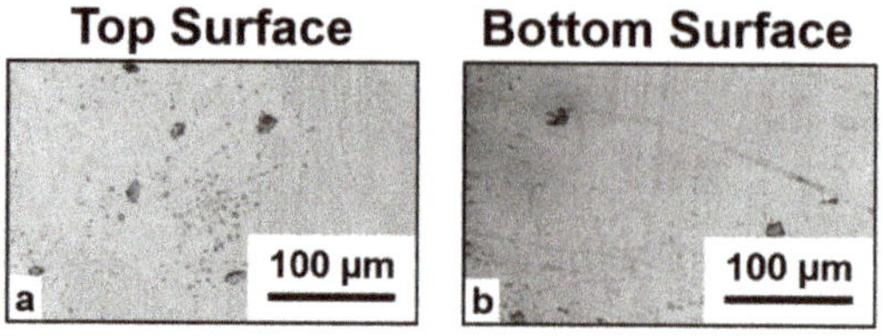

Top Surface Bottom Surface

a 100 μm b 100 μm

Figure 5. Top (**a**) and bottom (**b**) surface of the target before laser irradiation.

As the pulse energy was increased to above 25 μJ, a clear ablation occurred at the bottom of the sample. At an even higher value of pulse energy of 35 μJ, the damaged region merged with the laser trace, reaching an extension of almost 160 μm and resulting in cuts of poor quality.

In Figure 6, the 3D and line profiles of the cut edges obtained for 50 kHz of repetition rate, 1 mm/s of scan speed and pulse energies of 35 μJ and 20 μJ, respectively, are shown.

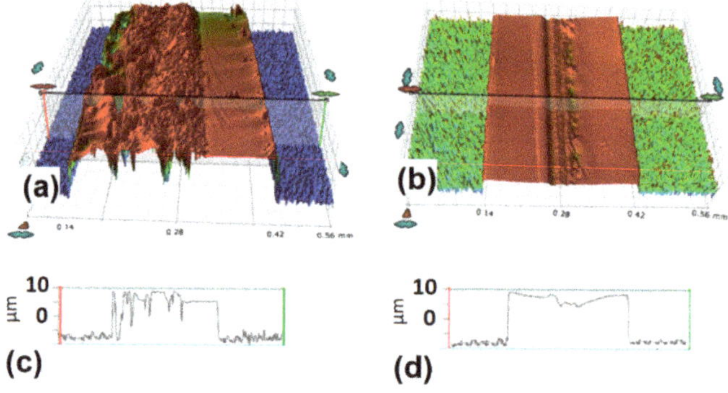

Figure 6. Three-dimensional profiles and line profiles of the cuts obtained at 50 kHz, scan speed of 1 mm/s and pulse energy of (**a**,**c**) 35 μJ and (**b**,**d**) 20 μJ, respectively.

At the higher pulse, energy evident laser-induced damage is present, as also clearly visible in Figure 4d,l. The cut edge obtained with the lower pulse energy has a significantly smoother surface, except for slight erosion lines on the right side and a small depression, only a few micrometers high, positioned at the laser trace in the middle of the cut edge.

4. Conclusions

We performed a systematic study on the stealth dicing process of quartz plates using ultrashort laser pulses. The influence of the main laser parameters as the repetition rate, the pulse energy, the translation speed and, as a consequence, the pulses overlap, on the cut efficiency and quality has been thoroughly investigated.

We have found that the single pass self-induced laser cutting of quartz is possible within the range of plate thickness explored in this work (250 μm). The physical mechanisms leading to the cleavage have been ascribed to the accumulation, pulse after pulse, of tensile stresses, generated by the rapid increase of temperature experienced in the focal volume after the nonlinear absorption of laser energy. Such stresses cause micro-fractures that produce the cut, while propagating along the laser path and throughout the entire thickness of the plate.

The repetition rate has been found to be a key parameter to generate a controlled and clean cut, since this parameter defines the time delay between successive pulses and thus the amount of stress accumulated in the focal volume. In fact, a repetition rate of 25 kHz was found to be too low to initiate the crack, while at 100 kHz, the bottom part of the plates was completely disrupted, due to the excessive laser induced stress. The optimal value of repetition rate, which has allowed obtaining a neat cut without any significant damage above or below the laser modified zone, was 50 kHz.

Increasing the overlap between pulses or the pulse energy and keeping the same repetition rate resulted in reducing the quality of the final cut, with the appearance of erosion lines and chipping around the laser modified zone. This indicates that a higher energy load causes, once again, excessive laser induced stresses and/or the initiation of ablation from the top or the bottom surface, thus making the stealth dicing process unstable.

Author Contributions: Conceptualization, C.G., A.V. and A.A.; methodology, C.G. and A.V.; formal analysis, C.G. and A.V.; investigation, C.G. and A.V.; data curation, C.G.; writing—original draft preparation, C.G and A.V.; visualization, C.G., writing—review and editing, C.G., A.V. and A.A.; supervision, A.A.; funding acquisition, A.A. All authors have read and agreed to the published version of the manuscript.

Funding: This research was funded by the Italian Ministry of Education, University and Research (MIUR) within the Project AIM184902B-1- ATT1.

Acknowledgments: The authors gratefully acknowledge the PolySense Lab (http://polysense.poliba.it/) for the provision of the quartz samples and Pietro Paolo Calabrese for his technical support.

Conflicts of Interest: The authors declare no conflict of interest.

References

1. Yadav, A.; Kbashi, H.; Kolpakov, S.; Gordon, N.; Zhou, K.; Rafailov, E.U. Stealth dicing of sapphire wafers with near infra-red femtosecond pulses. *Appl. Phys. A Mater. Sci. Process.* **2017**, *123*, 369. [CrossRef]
2. Trotta, G.; Volpe, A.; Ancona, A.; Fassi, I. Flexible micro manufacturing platform for the fabrication of PMMA microfluidic devices. *J. Manuf. Process.* **2018**, *35*, 107–117. [CrossRef]
3. Volpe, A.; Gaudiuso, C.; Ancona, A. Sorting of particles using inertial focusing and laminar vortex technology: A review. *Micromachines* **2019**, *10*, 594. [CrossRef]
4. Matsumaru, K.; Takata, A.; Ishizaki, K. Advanced thin dicing blade for sapphire substrate. *Sci. Technol. Adv. Mater.* **2005**, *6*, 120–122. [CrossRef]
5. Nisar, S.; Li, L.; Sheikh, M.A. Laser glass cutting techniques—A review. *J. Laser Appl.* **2013**, *25*, 042010. [CrossRef]
6. Zhimalov, A.B.; Solinov, V.F.; Kondratenko, V.S.; Kaplina, T.V. Laser cutting of float glass during production. *Glas. Ceram. (Engl. Transl. Steklo i Keram.)* **2006**, *63*, 319–321. [CrossRef]
7. Zhou, M.; Ngoi, B.K.A.; Yusoff, M.N.; Wang, X.J. Tool wear and surface finish in diamond cutting of optical glass. *J. Mater. Process. Technol.* **2006**, *174*, 29–33. [CrossRef]
8. Iliescu, C.; Chen, B.; Miao, J. On the wet etching of Pyrex glass. *Sens. Actuators A Phys.* **2008**, *143*, 154–161. [CrossRef]
9. Etching with Hydrofluoric Acid. Available online: https://www.microchemicals.com/technical_information/hf_etching.pdf (accessed on 21 March 2020).
10. 2004/394/EC. *COMMISSION RECOMMENDATION of 29 April 2004 on the Results of the Risk Evaluation and the Risk Reduction Strategies for the Substances: Acetonitrile; Acrylamide; Acrylonitrile; Acrylic Acid; Butadiene; Hydrogen Fluoride; Hydrogen Peroxide; Methacrylic Acid; Methyl Methacrylate; Toluene; Trichlorobenzene*; Official Journal of the European Union: Brussels, Belgium, 2004; pp. 72–122.
11. Available online: https://www.uab.edu/ehs/images/docs/chem/HFUserGuide-2016-09-29.pdf (accessed on 21 March 2020).
12. Jain, V.K.; Adhikary, S. On the mechanism of material removal in electrochemical spark machining of quartz under different polarity conditions. *J. Mater. Process. Technol.* **2008**, *200*, 460–470. [CrossRef]
13. Kuo, K.Y.; Wu, K.L.; Yang, C.K.; Yan, B.H. Wire electrochemical discharge machining (WECDM) of quartz glass with titrated electrolyte flow. *Int. J. Mach. Tools Manuf.* **2013**, *72*, 50–57. [CrossRef]
14. Nguyen, K.H.; Lee, P.A.; Kim, B.H. Experimental investigation of ECDM for fabricating micro structures of quartz. *Int. J. Precis. Eng. Manuf.* **2015**, *16*, 5–12. [CrossRef]
15. Wang, J.; Jia, Z.; Guo, Y.B. Shape-cutting of quartz glass by spark discharge-assisted diamond wire sawing. *J. Manuf. Process.* **2018**, *34*, 131–139. [CrossRef]

16. Choi, W.S.; Kim, J.H.; Kim, J. Thermal cleavage on glass by a laser-induced plume. *Opt. Lasers Eng.* **2014**, *53*, 60–68. [CrossRef]
17. Garibotti, D.J. Dicing of Micro-Semiconductors. U.S. Patent 3,112,850, 3 December 1963.
18. Serdyukov, A.N.; Shershnev, E.B.; Nikityuk, Y.V.; Sholokh, V.F.; Sokolov, S.I. Features of controlled laser thermal cleavage of crystal quartz. *Crystallogr. Rep.* **2012**, *57*, 792–797. [CrossRef]
19. Xu, J.; Hu, H.; Zhuang, C.; Ma, G.; Han, J.; Lei, Y. Controllable laser thermal cleavage of sapphire wafers. *Opt. Lasers Eng.* **2018**, *102*, 26–33. [CrossRef]
20. Tsai, C.H.; Liou, C.S. Fracture mechanism of laser cutting with controlled fracture. *J. Manuf. Sci. Eng. Trans. ASME* **2003**, *125*, 519–528. [CrossRef]
21. Cheng, X.; Yang, L.; Wang, M.; Cai, Y.; Wang, Y.; Ren, Z. The unbiased propagation mechanism in laser cutting silicon wafer with laser induced thermal-crack propagation. *Appl. Phys. A Mater. Sci. Process.* **2019**, *125*, 479. [CrossRef]
22. Pereles-Santiago, V.; Washington, M.; Brugan, P.; Cai, G.; Akarapu, R.; Pulford, S.; Segall, A.E. Faster and damage-reduced laser cutting of thick ceramics using a simultaneous prescore approach. *J. Laser Appl.* **2005**, *17*, 219–224. [CrossRef]
23. Bulychev, N.A.; Kuznetsova, E.L.; Rabinskiy, L.N.; Tushavina, O.V. Theoretical investigation of temperature-gradient induced glass cutting. *Nanosci. Technol. Int. J.* **2019**, *10*, 123–131. [CrossRef]
24. Russ, S.; Siebert, C.; Eppelt, U.; Hartmann, C.; Faißt, B.; Schulz, W. Picosecond laser ablation of transparent materials. In Proceedings of the SPIE 8608, San Francisco, CA, USA, 2–7 February 2013; p. 86080E.
25. Kumkar, M.; Bauer, L.; Russ, S.; Wendel, M.; Kleiner, J.; Grossmann, D.; Bergner, K.; Nolte, S. Comparison of different processes for separation of glass and crystals using ultrashort pulsed lasers. In Proceedings of the SPIE 8972, San Francisco, CA, USA, 1–6 February 2014; p. 897214.
26. Vanagas, E.; Kawai, J.; Tuzhilin, D.; Kudryashov, I.; Mizuyama, A.; Nakamura, K.G.; Kondo, K.I.; Koshihara, S.Y.; Takesada, M.; Matsuda, K.; et al. Glass cutting by femtosecond pulsed irradiation. *J. Microlithogr. Microfabr. Microsyst.* **2004**, *3*, 358–363. [CrossRef]
27. Hélie, D.; Vallée, R. Micromachining of thin glass plates with a femtosecond laser. In Proceedings of the SPIE, Quebec, QC, Canada, 4 August 2009; Volume 7386, p. 738639.
28. Tsai, W.-J.; Gu, C.-J.; Cheng, C.-W.; Horng, J.-B. Internal modification for cutting transparent glass using femtosecond Bessel beams. *Opt. Eng.* **2013**, *53*, 051503. [CrossRef]
29. Dudutis, J.; Stonys, R.; RaČiukaitis, G.; GeČys, P. Bessel beam asymmetry control for glass dicing applications. *Procedia CIRP* **2018**, *74*, 333–338. [CrossRef]
30. Lopez, J.; Mishchik, K.; Chassagne, B.; Javaux-Leger, C.; Hönninger, C.; Mottay, E.; Kling, R. Glass cutting using ultrashort pulsed bessel beams. In Proceedings of the International Congress on Applications of Laser & Electro-Optics (ICALEO), Georgia, GA, USA, 18–22 October 2015; Volume 60.
31. Danielyan, G.L.; Shilov, I.P.; Zamyatin, A.A.; Makovetskii, A.A.; Kochmarev, L.Y. Quartz Optical Fibers with Increased Content of Fluorine for Fluorimeter Probes. *J. Commun. Technol. Electron.* **2019**, *64*, 1123–1126. [CrossRef]
32. Elefante, A.; Giglio, M.; Sampaolo, A.; Menduni, G.; Patimisco, P.; Passaro, V.M.N.; Wu, H.; Rossmadl, H.; Mackowiak, V.; Cable, A.; et al. Dual-Gas Quartz-Enhanced Photoacoustic Sensor for Simultaneous Detection of Methane/Nitrous Oxide and Water Vapor. *Anal. Chem.* **2019**, *91*, 12866–12873. [CrossRef]
33. Boyd, R. *Nonlinear Optics*, 2nd ed.; Academic Press: San Diego, CA, USA, 2003; Chapter 7; pp. 311–323.
34. Wang, H.; Ji, X.-L.; Deng, Y.; Li, X.-Q.; Yu, H. Theory of the quasi-steady-state self-focusing of partially coherent light pulses in nonlinear media. *Opt. Lett.* **2020**, *45*, 710–713. [CrossRef]
35. Bergner, K.; Seyfarth, B.; Lammers, K.A.; Ullsperger, T.; Döring, S.; Heinrich, M.; Kumkar, M.; Flamm, D.; Tünnermann, A.; Nolte, S. Spatio-temporal analysis of glass volume processing using ultrashort laser pulses. *Appl. Opt.* **2018**, *57*, 4618–4632. [CrossRef]
36. Bulgakova, N.M.; Zhukov, V.P.; Mirza, I.; Meshcheryakov, Y.P.; Tomášťík, J.; Michálek, V.; Haderka, O.; Fekete, L.; Rubenchik, A.M.; Fedoruk, M.P.; et al. Ultrashort-pulse laser processing of transparent materials: Insight from numerical and semi-analytical models. In Proceedings of the SPIE 9735, San Francisco, CA, USA, 14 March 2016; p. 97350N.
37. Mishchik, K.; Beuton, R.; Dematteo Caulier, O.; Skupin, S.; Chimier, B.; Duchateau, G.; Chassagne, B.; Kling, R.; Hönninger, C.; Mottay, E.; et al. Improved laser glass cutting by spatio-temporal control of energy deposition using bursts of femtosecond pulses. *Opt. Express* **2017**, *25*, 33271. [CrossRef]

38. Sakakura, M.; Shimotsuma, Y.; Miura, K. Observation of stress wave and thermal stress in ultrashort pulse laser bulk processing inside glass. *J. Laser Micro Nanoeng.* **2017**, *12*, 159–164. [CrossRef]
39. Volpe, A.; Di Niso, F.; Gaudiuso, C.; De Rosa, A.; Vázquez, R.M.; Ancona, A.; Lugarà, P.M.; Osellame, R. Femtosecond fiber laser welding of PMMA. In Proceedings of the SPIE 9351, San Francisco, CA, USA, 12 March 2015; p. 935106.
40. Gaudiuso, C.; Giannuzzi, G.; Volpe, A.; Lugarà, P.M.; Choquet, I.; Ancona, A. Incubation during laser ablation with bursts of femtosecond pulses with picosecond delays. *Opt. Express* **2018**, *26*, 3801–3813. [CrossRef]
41. Tanvir Ahmmed, K.M.; Grambow, C.; Kietzig, A.M. Fabrication of micro/nano structures on metals by femtosecond laser micromachining. *Micromachines* **2014**, *5*, 1219–1253. [CrossRef]

Article

A Microfluidic Mixer of High Throughput Fabricated in Glass Using Femtosecond Laser Micromachining Combined with Glass Bonding

Jia Qi [1,2], Wenbo Li [1,2,3], Wei Chu [4,*], Jianping Yu [1,2,5], Miao Wu [4], Youting Liang [4], Difeng Yin [1,2], Peng Wang [1,2,5], Zhenhua Wang [4], Min Wang [4] and Ya Cheng [1,4,6,7,*]

1 State Key Laboratory of High Field Laser Physics, Shanghai Institute of Optics and Fine Mechanics, Chinese Academy of Sciences, Shanghai 201800, China; qijia@siom.ac.cn (J.Q.); liwb@shanghaitech.edu.cn (W.L.); 18722372031@163.com (J.Y.); yindf@siom.ac.cn (D.Y.); wangpeng2015@siom.ac.cn (P.W.)
2 University of Chinese Academy of Sciences, Beijing 100049, China
3 School of Physical Science and Technology, ShanghaiTech University, Shanghai 200031, China
4 XXL-The Extreme Optoelectromechanics Laboratory, School of Physics and Electronic Science, East China Normal University, Shanghai 200241, China; wumiao1993@126.com (M.W.); 15253172638@163.com (Y.L.); zhwang@phy.ecnu.edu.cn (Z.W.); mwang@phy.ecnu.edu.cn (M.W.)
5 School of Physics Science and Engineering, Tongji University, Shanghai 200092, China
6 State Key Laboratory of Precision Spectroscopy, East China Normal University, Shanghai 200062, China
7 Collaborative Innovation Center of Light Manipulations and Applications, Shandong Normal University, Jinan 250358, China
* Correspondence: wchu@phy.ecnu.edu.cn (W.C.); ya.cheng@siom.ac.cn (Y.C.)

Received: 28 December 2019; Accepted: 18 February 2020; Published: 19 February 2020

Abstract: We demonstrate a microfluidic mixer of high mixing efficiency in fused silica substrate using femtosecond laser-induced wet etching and hydroxide-catalysis bonding method. The micromixer has a three-dimensional geometry, enabling efficient mixing based on Baker's transformation principle. The cross-sectional area of the fabricated micromixer was 0.5×0.5 mm^2, enabling significantly promotion of the throughput of the micromixer. The performance of the fabricated micromixers was evaluated by mixing up blue and yellow ink solutions with a flow rate as high as 6 mL/min.

Keywords: ultrafast laser microfabrication; microfluidic; glass bonding

1. Introduction

Mixing is one of the dominating processes in chemical reactions and analyses. With microfluidic technology, various schemes have been conceived and implemented to realize highly efficient mixing of liquids by manipulating micro- and nanoscale fluids in sophisticated manners [1–6]. The geometries that have been incorporated into the microfluidic channels for promoting mixing efficiency include T-shaped microchannel, H-shaped micromixer, and grooved micromixer, etc. [7–12]. In particular, it has been demonstrated that a three-dimensional (3D) passive micromixer, which was designed basing on the Baker's transformation concept, can enable fast and efficient mixing even in the low-Reynolds-number condition [13]. Many planar manufacturing approaches, such as casting or injection molding, have been adopted to fabricate on-chip micromixer devices. However, these methods are inadequate for three dimension (3D) complex structures fabrication. 3D printing technology, which can be employed to produce 3D structures, always uses organic materials (i.e., epoxy resin). The organic materials are unsuitable for some microfluidic devices fabrication, since they are easily modified or dissolved by chemical reagents and damaged in high-temperature or high-pressure environments. In contrast, glass materials are chemically stable and resistant to corrosion, high temperature and pressure, making them excellent candidates for microfluidic chips preparation and functionalization. As a maskless technology, femtosecond laser direct writing (FLDW) enables rapid prototyping and provides a straightforward approach to fabricate 3D structures inside photosensitive materials, including polymer and glass. The capability of 3D prototyping with high resolution in a wild range of transparent materials makes FLDW a promising and superior technology for the fabrication of microfluidic devices. It should be mentioned that the 3D micromixer was fabricated using FLDW of glass [14–18], which has been proved to be a straightforward approach of fabricating 3D microfluidic structures and integrated optofluidic devices [19–22].

The femtosecond laser micromachining can be an ideal tool for fabricating such structures owing to its high fabrication precision and 3D capacity. However, the micrometer scale mixers suffer from a relatively low throughput for various kinds of chemical reactions. The solution is to fabricate microfluidic mixers consisting of relatively thicker and longer channels, which is nevertheless challenging for the current state-of-the-art femtosecond laser-induced selective etching (FLSIE) technique [23].

Here, we demonstrate the fabrication of 3D micromixers of large footprint sizes in glass using femtosecond laser micromachining. We improve the fabrication efficiency by optimizing the laser pulse duration. The 3D micromixer, designed basing on the Baker's transformation, is constructed by bonding two substrates with complementary microfluidic channels fabricated on the surface. The advantage of our design is that for the upper and lower halves of the mixer, they both have a 2D planar geometry, which facilitates obtaining the designed structures by FLSIE technique, while the mixer produced after bonding has a true 3D geometry to implant the Baker's transformation operation. The elimination of any vertical structures in the upper and lower halves of the micromixer provides more tolerance for the alignment between the upper and lower halves during the bonding; while the employment of Baker's transformation for the mixing leads to high mixing efficiency regardless of the flow rates in the microchannels. Compared to the previous micromixers designed and achieved basing on Baker's transformation, the throughput of our micromixer can be significantly enhanced. As shown in Table 1, we compared the performances of different micromixer devices in detail.

Table 1. Comparison of three different types of micromixer devices designed and fabricated basing on the Baker's transformation.

Channel Width × Height	Maximum Flow Rate	Materials	Processing Method	Ref.
160 μm × 20 μm	~200 μL/min	PDMS	PDMS molds	[24]
50 μm × 75 μm	~20 μL/min	Porous glass	Femtosecond laser direct writing and annealing	[14]
500 μm × 500 μm	6 mL/min	Fused silica	FLISE and bonding	This work

2. Device Design and Numerical Simulations of Mixing Process

Figure 1a illustrates the mixing effect in a single mixing unit. The unit can be arranged into a chain for the construction of a high efficiency and high throughput microfluidic mixer. The mixing unit features a 3D microstructure which splits, routes, and reorganizes the fluidic streams within the upper and lower halves of the channel into an array of alternatively arranged sub-streams, which is the so-called Baker's transformation. In such a manner, the number of microfluidic streams can be quadrupled after they pass each of the mixing unit. The working mechanism provides an efficient way of mixing with the relatively simple geometry as compared to that in [13,14]. The design of the whole micromixer is schematically illustrated in Figure 1b. First, the upper half of the micromixer is engraved into the fused silica substrate using femtosecond laser-assisted chemical wet etching. Likewise, the lower half of the micromixer is fabricated using the same technique. The two substrates are finally bonded into the micromixer using a hydroxide-catalysis bonding technique as described later in this paper.

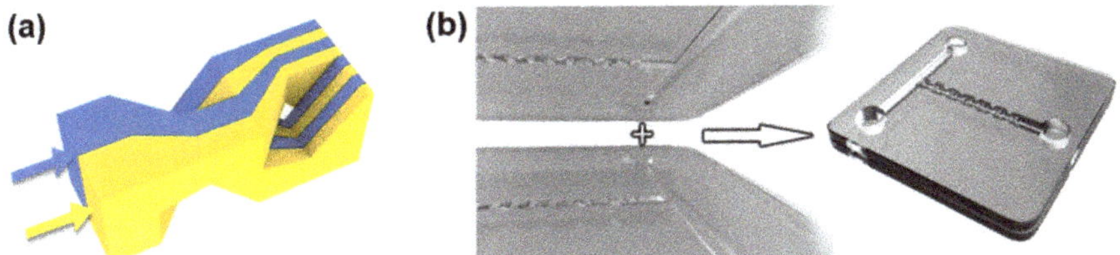

Figure 1. (**a**) Schematic view of working mechanism of the micromixer. The two microfluidic streams sent into the mixing unit are divided into four sub-streams alternatively spaced with each other at the middle of the unit and further divided into eight streams at the exit; (**b**) Schematic of the 3D micromixer constructed by bonding two substrates with microfluidic channels fabricated by femtosecond laser micromachining on the top surfaces.

The mixing performance of the designed 3D micromixer composed of six mixing units is numerically simulated by solving the microfluidic incompressible Navier-Stokes and convection diffusion equations using a finite element analysis software (COMSOL Multiphysics 5.4, COMSOL Multiphysics GmbH, Göttingen, Germany). The simulations results compared with that of a straight microfluidic channel are illustrated in Figure 2. The two structures are of the same cross-sectional size and total length. One can see that in the 1D straight channel in Figure 2a, mixing only occurs at the interface of two streams as a result of diffusion. Owing to the laminar flow, which dominates at low Renolde numbers in the microfluidic channels, the overall mixing efficiency is low. In contrast, the 3D mixer in Figure 2b shows an excellent mixing effect thanks to the working mechanism described above.

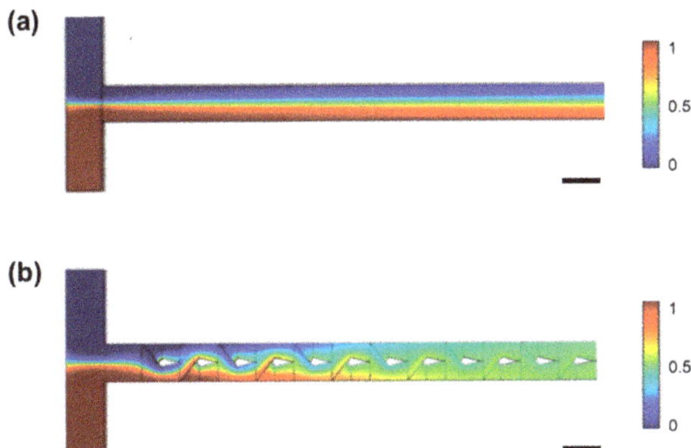

Figure 2. Numerical simulations of mixing performances at a flow rate of 2 mL/min in (**a**) a T-shape straight microchannel with a rectangular cross section and (**b**) a 3D micromixer consisting of six mixing units. The straight microchannel and the 3D micromixer have the same total length and the same cross section area. Scale bar in (**a**,**b**): 0.5 mm.

3. Fabrication of the 3D Micromixers

The 3D micromixer is fabricated using femtosecond laser-assisted chemical wet etching technique and the hydroxide-catalysis bonding method. First, the upper and the lower half of the micromixer is engraved into the fused silica substrate using femtosecond laser-assisted chemical wet etching. The femtosecond laser pulses (1030 nm, up to 400 μJ, 270 fs) were provided by a commercial femtosecond laser source (Pharos, Light Conversion Ltd., Vilnius, Lithuania). The duration of the laser pulse can be tuned from 270 fs to 15 ps by adjusting the distance between the gratings in compressor. After passing through an attenuator and a beam expanding system, the laser pulses were then focused into the fused silica glass using an objective lens (Olympus MPLFLN, 10×, NA = 0.3, OLYMPUS, Tokyo, Japan). A motion stage (ANT130-110-L-ZS, Aerotech Inc., Pittsburgh, PA, USA) was used to translate the objective lens along Z direction to control the depth of the focus, and the fused silica glass sample was mounted on an XY motion stage (ABL15020WB and ABL15020, Aerotech Inc., Pittsburgh, PA, USA) and smoothly translated with a positioning precision of 100 nm. Both the translation stages were controlled using a high-performance motion controller (A3200, Aerotech Inc., Pittsburgh, PA, USA). In our fabrication, the repetition rate of the laser was set to 100 kHz, and the laser pulse duration was set as 4 ps [25,26]. The laser focal spot was scanned along the pre-designed paths layer by layer with a layer spacing of 10 μm to produce the microchannels on both glass substrates. The scan process was performed from the bottom to the top of the substrate, and the scan speed was fixed at 10 mm/s.

After laser irradiation, the glass samples were immersed in a solution of potassium hydroxide (KOH) with a concentration of 10 mol/L to selectively remove the glass material irradiated by the laser pulses. The microchannels on both glass substrates can form after the etching in KOH solution.

Lastly, the two substrates were combined into the micromixer using the hydroxide-catalysis bonding method. First, the top surfaces of the two glass substrates were polished, and the two substrates were ultrasonically cleaned in acetone for 10 min and subsequently in distilled water for 10 min. Then, a drop of 2% sodium hydroxide (2% NaOH) solution was applied to the bonding surfaces of the two glass substrates. Afterwards, we carefully adjusted the position of the two glass substrates under the transmission illumination microscope to ensure an accurate alignment between the microchannels engraved in the two substrates. After that, the whole sample was squeezed gently, held for 24 h at the room temperature, and then annealed at 200 °C for another 24 h to reinforce the bonding strength [27,28]. The procedures of the micromixer fabrication are illustrated in Figure 3.

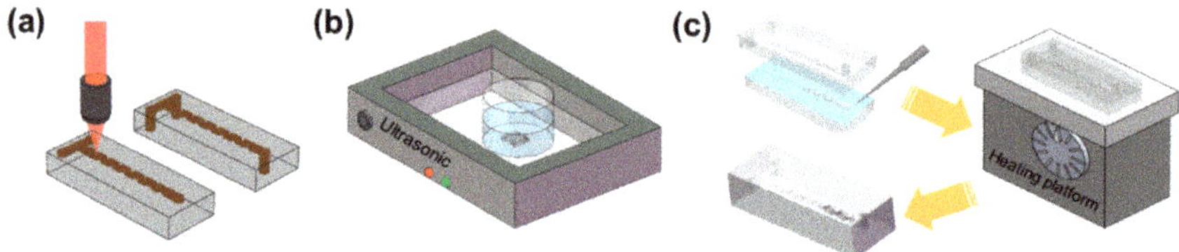

Figure 3. Fabrication procedures: (**a**) ultrafast laser direct writing; (**b**) chemical wet etching; and (**c**) hydroxide-catalysis bonding.

4. Results and Discussion

The top view micrograph of the fabricated 3D microfluidic mixer is shown in Figure 4a. One can see that it contains six mixing units. The sharpness of the edges and corners in the fabricated structure provide the evidence of the high machining quality of the femtosecond laser. Figure 4b presents the detailed top view image of one of the mixing units. Figure 4c presents the top view image of half of the micromixer before bonding. Figure 4d,e exhibit the 3D profiles of the structure in Figure 4c from different angles of view, showing a maximum depth of ~270 μm in the fabricated microchannel. The optical micrograph of the cross section was illustrated in the inset of Figure 4e. The cross section shows a square profile with a side length of 0.5 mm. Because of the relatively large height and width of the micromixer, the production throughput can be efficiently promoted, which is highly in demand by industrial application. It should be noted that the measured roughness of the micromixer's inner surface in the region of 200 μm × 200 μm is 872 nm, which is orders of magnitude smaller when compared to the size of the micromixer. As a consequence, the impact of the surface roughness to the mixing process can be ignored.

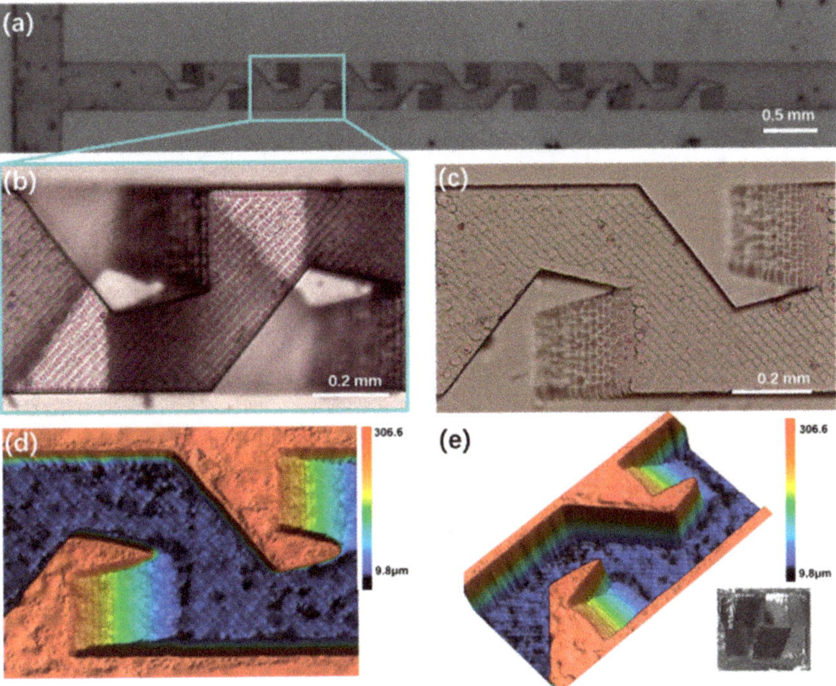

Figure 4. Top view optical micrograph of (**a**) fabricated micromixers; (**b**) the detailed features of mixing unit; (**c**) the microchannel on half glass substrate before bonding. (**d**,**e**) are the 3D images of (**c**) from different view angles captured by laser confocal microscopy. Inset: Image of the cross section at the end of the mixing unit after bonding.

At last, we experimentally demonstrated the mixing of the fabricated 3D micromixer using two kinds of ink of blue and yellow colors, as shown in Figure 5. The experimental performances of the 3D micromixer compared with that of the straight microchannels were conducted at three different flow rates of 1 mL/min (Figure 5a,b), 2 mL/min (Figure 5c,d), and 6 mL/min (Figure 5e,f), which correspond to Reynolds numbers of 8.33, 16.67, and 50 in the micromixer respectively. It can be seen that in the straight channel, the mixing efficiency contributed by diffusion decreases to the increasing flow rate. Overall, the mixing effect in Figure 5a–c are much weaker than that in the 3D mixer as shown in Figure 5b,d,f respectively. Interestingly, as the mixing in the 3D micromixer is achieved dominatingly by the Baker's transformation mechanism but not by diffusion, the mixing performances at different flow rates appear similar to each other. The stabilization of mixing efficiency at the variable flow rate provides a controllable way to manipulate chemical/biological reactions in the microchannel by only changing the flow rate, which leads to a predictable change of the dwell time of the reactants in the microchannel.

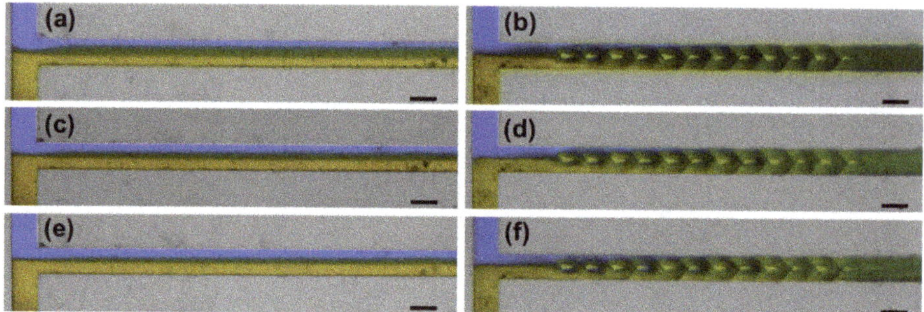

Figure 5. Microscope images of the mixing behaviors of the blue and yellow ink solutions in straight channel at a flow rate of (**a**) 1 mL/min, (**c**) 2 mL/min and (**e**) 6 mL/min, and in fabricated 3D micromixer at a flow rate of (**b**) 1 mL/min, (**d**) 2 mL/min and (**f**) 6 mL/min. Scale bar in (**a**–**f**): 0.5 mm.

5. Conclusions

To conclude, we have designed and fabricated a 3D microfluidic mixer based on femtosecond laser micromachining technology and hydroxide-catalysis bonding method. Both our simulation and experimental results show that the device can realize efficient microfluidic mixing. The compact and efficient 3D micromixer can be used in applications ranging from chemical/biological analysis and microfluidic synthesis of materials to fine chemistry microreaction. Our technique also has promising potential in electrophoretic and some other relative applications [29] because it can offer the capabilities of 3D fabrication, controllable fluidic throughput and multifunctional integration.

Author Contributions: Conceptualization, Y.C.; methodology, Y.C., W.C., J.Q., W.L. and P.W.; software, W.C., J.Q. and P.W.; validation, J.Q., W.L.; formal analysis, J.Q., W.L.; simulations, J.Q., W.L. and M.W. (Min Wang); investigation, J.Q., P.W., D.Y., and Y.L.; resources, J.Q., P.W., W.C., J.Y., M.W. (Miao Wu), W.L. and Z.W.; data curation, J.Q., W.C.; writing—original draft preparation, Y.C., W.C. and J.Q.; writing—review and editing, Y.C., W.C. and J.Q.; visualization, J.Q. and J.Y.; supervision, Y.C.; funding acquisition, Y.C., and W.C. All authors have read and agreed to the published version of the manuscript.

Funding: The work is supported by Key Project of the Shanghai Science and Technology Committee (No. 18DZ1112700).

Conflicts of Interest: The authors declare no conflict of interest.

References

1. Losey, M.W.; Schmidt, M.A.; Jensen, K.F. Microfabricated Multiphase Packed-Bed Reactors: Characterization of Mass Transfer and Reactions. *Ind. Eng. Chem. Res.* **2001**, *40*, 2555–2562. [CrossRef]
2. Wu, N.; Wu, S.-Z.; Xu, J.; Niu, L.-G.; Midorikawa, K.; Sugioka, K. Hybrid femtosecond laser microfabrication to achieve true 3D glass/polymer composite biochips with multiscale features and high performance: The concept of ship-in-a-bottle biochip. *Laser Photon. Rev.* **2014**, *8*, 458–467. [CrossRef]
3. Lee, C.-Y.; Chang, C.-L.; Wang, Y.-N.; Fu, L.-M. Microfluidic Mixing: A Review. *Int. J. Mol. Sci.* **2011**, *12*, 3263–3287. [CrossRef]
4. Xia, H.; Wang, Z.P.; Koh, Y.X.; May, K.T. A microfluidic mixer with self-excited 'turbulent' fluid motion for wide viscosity ratio applications. *Lab Chip* **2010**, *10*, 1712–1716. [CrossRef] [PubMed]
5. Martin, L.; Meier, M.; Lyons, S.; Sit, R.V.; Marzluff, W.F.; Quake, S.R.; Chang, H.Y. Systematic reconstruction of RNA functional motifs with high-throughput microfluidics. *Nat. Methods* **2012**, *9*, 1192–1194. [CrossRef] [PubMed]
6. Gervais, L.; De Rooij, N.; Delamarche, E. Microfluidic Chips for Point-of-Care Immunodiagnostics. *Adv. Mater.* **2011**, *23*, H151–H176. [CrossRef]

7. Wang, W.; Zhao, S.; Shao, T.; Jin, Y.; Cheng, Y. Visualization of micro-scale mixing in miscible liquids using μ-LIF technique and drug nano-particle preparation in T-shaped micro-channels. *Chem. Eng. J.* **2012**, *192*, 252–261. [CrossRef]
8. Nimafar, M.; Viktorov, V.; Martinelli, M. Experimental comparative mixing performance of passive micromixers with H-shaped sub-channels. *Chem. Eng. Sci.* **2012**, *76*, 37–44. [CrossRef]
9. Cortes-Quiroz, C.A.; Azarbadegan, A.; Zangeneh, M.; Goto, A. Analysis and multi-criteria design optimization of geometric characteristics of grooved micromixer. *Chem. Eng. J.* **2010**, *160*, 852–864. [CrossRef]
10. Du, Y.; Zhang, Z.; Yim, C.; Lin, M.; Cao, X. Evaluation of Floor-grooved Micromixers using Concentration-channel Length Profiles. *Micromachines* **2010**, *1*, 19–33. [CrossRef]
11. Therriault, D.; White, S.; Lewis, J.A. Chaotic mixing in three-dimensional microvascular networks fabricated by direct-write assembly. *Nat. Mater.* **2003**, *2*, 265–271. [CrossRef] [PubMed]
12. Burns, M.A.; Johnson, B.N.; Brahmasandra, S.N.; Handique, K.; Webster, J.R.; Krishnan, M.; Sammarco, T.S.; Man, P.M.; Jones, D.; Heldsinger, D.; et al. An Integrated Nanoliter DNA Analysis Device. *Science* **1998**, *282*, 484–487. [CrossRef] [PubMed]
13. Carrière, P. On a three-dimensional implementation of the baker's transformation. *Phys. Fluids* **2007**, *19*, 118110. [CrossRef]
14. Liao, Y.; Song, J.; Li, E.; Luo, Y.; Shen, Y.; Chen, D.; Cheng, Y.; Xu, Z.; Sugioka, K.; Midorikawa, K. Rapid prototyping of three-dimensional microfluidic mixers in glass by femtosecond laser direct writing. *Lab Chip* **2012**, *12*, 746–749. [CrossRef] [PubMed]
15. Osellame, R.; Cerullo, G.; Ramponi, R. *Femtosecond Laser Micromachining: Photonic and Microfluidic Devices in Transparent Materials*; Springer: Berlin, Germany, 2012; Volume 123.
16. Kirby, B.J. *Micro-and Nanoscale Fluid Mechanics: Transport in Microfluidic Devices*; Cambridge University Press: Cambridge, UK, 2010.
17. Weisgrab, G.; Ovsianikov, A.; Costa, P. Functional 3D Printing for Microfluidic Chips. *Adv. Mater. Technol.* **2019**, *4*, 1900275. [CrossRef]
18. Au, A.K.; Huynh, W.; Horowitz, L.F.; Folch, A. 3D-Printed Microfluidics. *Angew. Chem. Int. Ed.* **2016**, *55*, 3862–3881. [CrossRef]
19. Cheng, Y.; Sugioka, K.; Midorikawa, K.; Masuda, M.; Toyoda, K.; Kawachi, M.; Shihoyama, K. Three-dimensional micro-optical components embedded in photosensitive glass by a femtosecond laser. *Opt. Lett.* **2003**, *28*, 1144–1146. [CrossRef]
20. Sugioka, K.; Cheng, Y. Femtosecond laser processing for optofluidic fabrication. *Lab Chip* **2012**, *12*, 3576. [CrossRef]
21. Hunt, H.C.; Wilkinson, J. Optofluidic integration for microanalysis. *Microfluid. Nanofluidics* **2007**, *4*, 53–79. [CrossRef]
22. Fan, X.; White, I.M. Optofluidic microsystems for chemical and biological analysis. *Nat. Photon.* **2011**, *5*, 591–597. [CrossRef]
23. Cheng, Y. Internal Laser Writing of High-Aspect-Ratio Microfluidic Structures in Silicate Glasses for Lab-on-a-Chip Applications. *Micromachines* **2017**, *8*, 59. [CrossRef]
24. Yasui, T.; Omoto, Y.; Osato, K.; Kaji, N.; Suzuki, N.; Naito, T.; Watanabe, M.; Okamoto, Y.; Tokeshi, M.; Shamoto, E.; et al. Microfluidic baker's transformation device for three-dimensional rapid mixing. *Lab Chip* **2011**, *11*, 3356. [CrossRef] [PubMed]
25. Wang, P.; Chu, W.; Li, W.; Tan, Y.; Liu, F.; Wang, M.; Qi, J.; Lin, J.; Zhang, F.; Wang, Z.; et al. Three-Dimensional Laser Printing of Macro-Scale Glass Objects at a Micro-Scale Resolution. *Micromachines* **2019**, *10*, 565. [CrossRef] [PubMed]
26. Li, X.; Xu, J.; Lin, Z.; Qi, J.; Wang, P.; Chu, W.; Fang, Z.; Wang, Z.; Chai, Z.; Cheng, Y. Polarization-insensitive space-selective etching in fused silica induced by picosecond laser irradiation. *Appl. Surf. Sci.* **2019**, *485*, 188–193. [CrossRef]
27. Rowan, S.; Twyford, S.; Hough, J.; Gwo, D.-H.; Route, R. Mechanical losses associated with the technique of hydroxide-catalysis bonding of fused silica. *Phys. Lett. A* **1998**, *246*, 471–478. [CrossRef]

28. Elliffe, E.J.; Bogenstahl, J.; Deshpande, A.; Hough, J.; Killow, C.; Reid, S.; Robertson, D.; Rowan, S.; Ward, H.; Cagnoli, G. Hydroxide-catalysis bonding for stable optical systems for space. *Class. Quantum Gravity* **2005**, *22*, S257–S267. [CrossRef]
29. Sun, K.; Suzuki, N.; Li, Z.; Araki, R.; Ueno, K.; Juodkazis, S.; Abe, M.; Noji, S.; Misawa, H. High-fidelity fractionation of ssDNA fragments differing in size by one-base on a spiral-channel electrophoretic chip. *Electrophoresis* **2009**, *30*, 4277–4284. [CrossRef]

MDPI
St. Alban-Anlage 66
4052 Basel
Switzerland
Tel. +41 61 683 77 34
Fax +41 61 302 89 18
www.mdpi.com

Micromachines Editorial Office
E-mail: micromachines@mdpi.com
www.mdpi.com/journal/micromachines

www.ingramcontent.com/pod-product-compliance
Lightning Source LLC
LaVergne TN
LVHW070707170726
843515LV00004B/505

* 9 7 8 3 0 3 6 5 3 3 1 0 0 *